Format Your Book WITH VELLUM

Format Your Book WITH VELLUM

A simplified Vellum tutorial with notes for PC users

Jody E. Skinner
Smidgen Press

Editing: Rogena Mitchell-Jones, RMJ Editing and Manuscript Service, www.rogenamitchell.com

Cover: Todd Damotte, www.damotte.com

ISBN | Paperback: 978-1-950536-00-9 | Hardcover: 978-1-950536-01-6

Compiled with Adobe InDesign

Created with Vellum

For my love,

without whom this project would still not be done.

Your encouragement and patience bless my life.

TYFYL

CONTENTS

INTRODUCTION

If you're reading this, you're probably in one of three categories:

1. I'm trying to "drive" Vellum and I want an instruction book.
2. I've heard of Vellum, but I'm not sure if it's for me.
3. What's Vellum? I just want to format my book.

Well, #2 and #3, welcome, and have a look around. There are several things that won't make complete sense without actually using the Vellum app—so while you're deciding if you want to actually take a test-drive, so to speak, you might find the scattered publishing tips useful. Thanks for stopping by (and feel free to stick around).

Category #1, though—you want an instruction book?

Format Your Book with Vellum is for YOU.

I wished for *years* for a book with start-to-finish tips and a few tricks. Something simple and straightforward with actual pages and some room to make notes. (And something that would show me what those crazy Vellum print options look like at actual size!)

So since I couldn't find what I wanted, I wrote it.

And since me-from-2017 isn't likely to benefit from this how-to collection, maybe it can help you, instead.

So break out the highlighter. Mark things up. (Maybe not a library copy. But you knew that already.) And I truly hope this tool will help you successfully format your own book with Vellum.

All the best!
~Jody E. Skinner

PART ONE

FORMAT YOUR BOOK FROM START TO FINISH

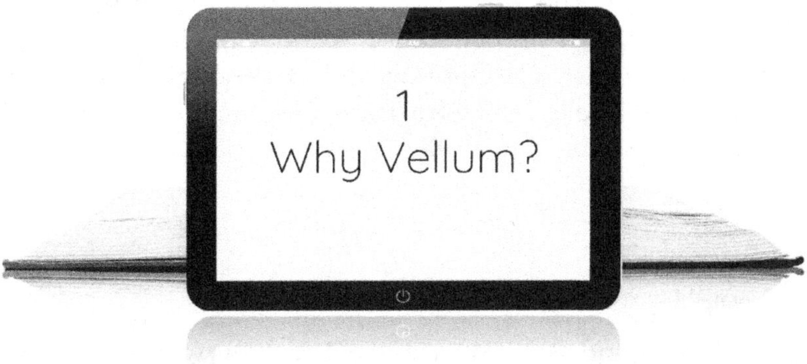

The universe called *self-publishing* is more expansive than most writers imagine. After all, can't anyone write a book and publish it online now? Well, yes, but there are standards to be met and requirements to follow.

Take ebooks, for example. For my own clients, they were pleased with the print formats I delivered, but I was embarrassed that my limited coding knowledge made the ebook files I provided look cheap by comparison. Ebooks *must* be less complex than print files, but they shouldn't deter or frustrate readers.

And paperbacks. Many writers would like to format their own book—how hard can it be?—but they need to set up page numbers correctly, embed fonts, insert images in line with text at 300 ppi/dpi (sized large enough for print but small enough for an ebook, by the way) . . . And those dratted page headers. Which pages have them? Which ones shouldn't, or the book will scream "homemade"? Ack!

THE CONUNDRUM

Amazon wasn't the first to create an ebook reader, but their launch of the Kindle in 2007, followed by Barnes & Noble's release of the Nook in 2009, paved the way for literally any writer to release an ebook. Many writers have done exactly that—but even thirteen years later, there are still glitches. Random characters, strange text breaks, or irregular font sizes are not uncommon. And sometimes a writer just says, "That's not what I wanted it to look like," when an ebook is finally up for sale.

In general, without knowing HTML, XML, or CSS—coding talk for the building blocks of an ebook—or without paying someone who does, there hasn't been a straightforward, dependable way to prevent ebook errors. So ebooks can be frustrating for readers . . . and difficult (or expensive) for everyday writers to fix.

Enter Vellum.

A SOLUTION

Brad Andalman and Brad West met when they worked together at Pixar Animation Studios. After they both left in 2012, they compared notes and recognized the need for *quality* ebooks. With their animation and software development experience, they realized if formatting ebooks was challenging for the general public, then they could offer a solution.

So "The Brads" launched the 180g company. They designed a software program with the proper coding and added enough design elements to be aesthetically pleasing. After testing repeatedly to ensure product compatibility with the ever-changing digital publishing world, they made Vellum available to the Mac world. Voilà! Nearly bulletproof ebooks.

Then, with the ease of formatting ebooks, the next thing users requested was print book formatting. The Brads worked through months of testing to be sure that a Vellum-created print file would pass the same quality tests as a Vellum ebook. In 2017, the paperback option was launched. The Brads listen to their users, keeping a list of every request and implementing new features as demand and feasibility allow.

BUT WILL VELLUM BE AVAILABLE FOR PC?

One of the most common questions is whether Vellum will be offered for PC along with Mac. According to the company, they are not working on a PC version, and have chosen instead to focus current development on providing new features and consistent updates for the Mac app.

But PC users, *there IS a PC workaround:* **MacinCloud**.

MacinCloud provides a connection to remote Mac servers around the world, allowing users to access Vellum from the comfort of their own PC. Vellum is pre-installed on MacinCloud servers, so if you choose to use their service, you can try out all of Vellum's features as soon as you log in. (See the "MacinCloud" chapter for more details.)

The Vellum app download is free, and there are no restrictions on Vellum's features for testing purposes. You will need to purchase a license when you want to generate or export files. At present, you can purchase an ebook-only version for $199 (U.S.) or ebook plus paperback for $249—a decent investment but one worth considering, especially if you will be putting out more than one book.

CAN WE FORMAT ANY BOOK WITH VELLUM?

Actually—yes and no. While Vellum can format books in just about any genre, some projects are too complex for Vellum to handle appropriately. Whether Vellum is a good fit depends on the type of project and the level of detail involved.

In fiction works, for example, a clean and straightforward manuscript would work well in Vellum, whereas a book with various custom fonts isn't feasible. There are ways to add personal touches at the start of a chapter or illustrations within the book, but there are no built-in options for things like layering text on images or adding watermarks.

A common lament among fiction writers has been how full-page illustrations were constrained by the set page margins, meaning authors with large maps or charts had to turn elsewhere to create their book or edit the final PDF. But, once again, The Brads listened to their users! As this book was heading to the printer, Vellum announced the ability to add full-page, full-bleed images and image spreads across two pages—or both. Find details at help.vellum.pub/elements/full-page-image.

Basic nonfiction works well in Vellum, but files with multiple levels of subheadings will require manual alignments and adjustments. Vellum offers capability for endnotes but not footnotes. Bulleted or numbered lists are limited to a single level—no sub-lists. Books with a complex structure are better served by programs like Calibre, Jutoh, or Sigil for ebooks, and Adobe InDesign or Affinity Publisher for print..

If Vellum's pros outweigh the cons, and you *just want to format your own book*, then I'd say you're ready for the power of being your own designer and making your book come to life. After all, that's the goal —right?

HOW TO USE THIS BOOK

Part I of *Format Your Book with Vellum*, as you might have guessed, is a start-to-finish formatting guide. If you've never used Vellum before, or you just want to see if you've been missing any steps, Part I will walk you through a full book project. I've left room for notes after every chapter, and I've scattered a few publishing and formatting tips for you, but some requirements may vary between online sales platforms.

Disclosure: While these are all my own notes, descriptions, and screenshots, all the basic Vellum how-to information can be found at help.vellum.pub—so this book should not be seen as a replacement for what the 180g team has so thoroughly provided. If you're happiest solving a puzzle all on your own, then Part I isn't something you need—and you probably aren't reading this book! Between added insights from the perspective of a PC user and my favorite Vellum tips, this book is written for someone who doesn't want to flip back and forth between browser tabs and prefers information collected in one place. (You know who you are!)

Part II in this book is unique. While every page and text sample in this book is generated in Vellum, I have used the magic of Adobe InDesign to compile the book, tab the page edges for quick reference, and to assemble Part II from over 180 separate files.

In Part II you'll find the following:

- **Appendix A**: Every chapter opener style as it appears in print
- **Appendix B**: All nine default font styles (plus the only "bonus" Vellum font) at all seven size settings
- **Appendix C**: All font styles in large print sizes

The font sizes shown in Appendix B and C are specific to the size of this book (6″x9″). I queried Vellum's help desk about font sizes, and this was the reply:

"Vellum's algorithm uses the selected *trim size* [in this book, 6″x9″] to determine the appropriate *font size*. While Vellum does not allow authors to select a particular point size (e.g. 12pt), they can use the Print Settings dialog to increase or decrease the font size. This ensures a high-quality book and leaves less room for formatting errors."

TROUBLESHOOTING

Whether I have needed a quick refresher course or hunted for a particular Vellum feature, I have kept the Vellum website's help section bookmarked (help.vellum.pub, as I mentioned). Here, there are a number of common topics and troubleshooting tips, and at the bottom of the page, there is also a Contact option to email Vellum's help desk directly, if needed.

For general Mac issues, operating system updates, etc., I frequent a Facebook group called Vellum Users. I scroll through or use the "search" feature to see if my question has already been asked, and whenever I have posted a question, the group has been helpful and supportive. Between this group of indie authors on Vellum Users and the Vellum Help website links, I've never had a question go unanswered.

LET'S GET GOING!

Visual learners: I've included loads of images (107 of them!) so you can skim and start clicking. (Off you go!)

Read-every-word folks: My directions will be slightly wordy, as I am assuming that some readers are new both to Mac *and* Vellum. Thanks for being gracious and skimming—or skipping—where appropriate.

New Mac users: Don't know what I mean by a Dock or a Launchpad? Start with the "New to Mac?" chapter for the essential commands you'll use to navigate Vellum and your new computer setup. You can also visit the "MacinCloud" and "Dropbox" chapters if you need help getting those off the ground. Then come back and start with the "Vellum Tour" chapter.

One last comment: Though this is a basic start-to-finish action plan, questions will arise that I haven't covered for the sake of simplicity and efficiency. For more information online, visit the Vellum tutorials (help.vellum.pub/tutorial) or find the Vellum Users group on Facebook. Also, Chuck Heintzelman's *The Author's Guide to Vellum* is a hefty book packed with information on every Vellum menu option, plus extra document prep and publishing tips.

Publishing a book is challenging, but it shouldn't be impossible. My sincere wish is that these formatting and self-publishing tips will be a source of hope and help to you.

Welcome to Vellum.

Created with Vellum version 2.6.3 | Book style: Chroma | Font: Palatino
Vellum files compiled/tabbed and page footers added using Adobe InDesign

---

Get your free Vellum Project Tracker, where you can collect your project details all in one place (and make notes as we go!). Visit smidgenpress.pub / tracker for more details.

Let's make a BOOK! By the time you complete the following chapters, you will have the tools to use your own Vellum license to generate ebook files and PDFs for print. First, let's get acquainted with the main controls in the Vellum interface.

A QUICK VELLUM TOUR

Contents/Text Editor. Vellum opens up to the Contents menu by default, which is connected to the Text Editor pane (*Fig. 1*). Here, you can make text adjustments and add or format items.

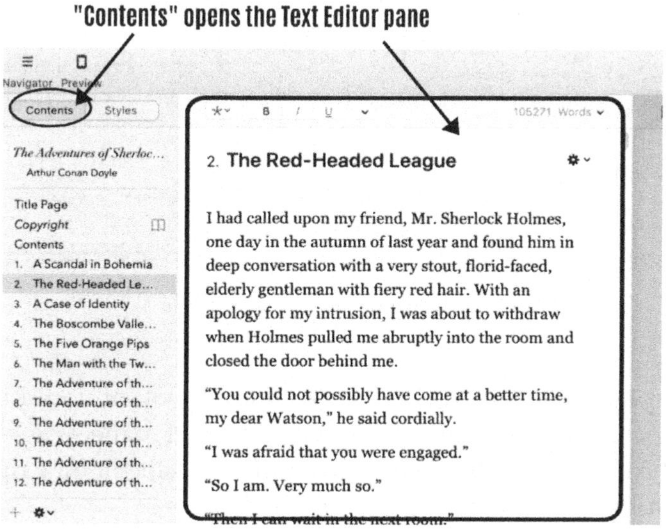

Fig. 1: Contents button and Text Editor pane

Navigator. In the Contents menu view, the Navigator pane shows all your book elements (*Fig. 2*). This can be turned off and on if you wish—just click the Navigator button at the top left.

Preview. The right side of the Vellum window is the Preview pane, where you can see your book text in different formats. This can be turned off or on by clicking the Preview icon, top left (*Fig. 3*).

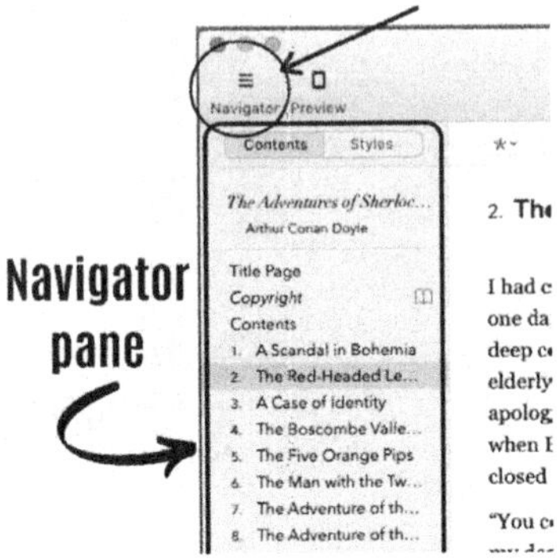

Fig. 2: The Navigator Pane

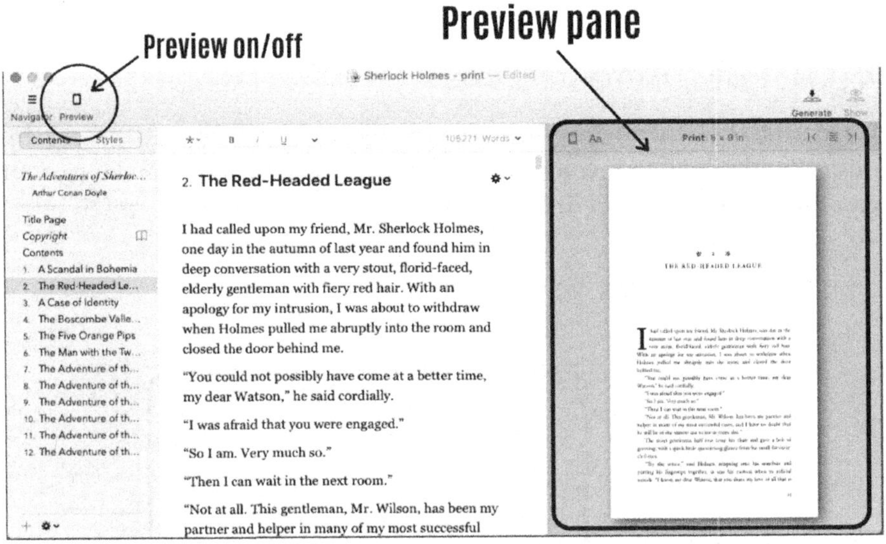

Fig. 3: Preview pane and on/off toggle

Styles/Style Carousel. To change the overall look of your book, click the Styles button to open the Style Carousel pane (*Fig. 4*). Select different categories in the Navigator pane and click left or right in the Style Carousel to view the appearance options.

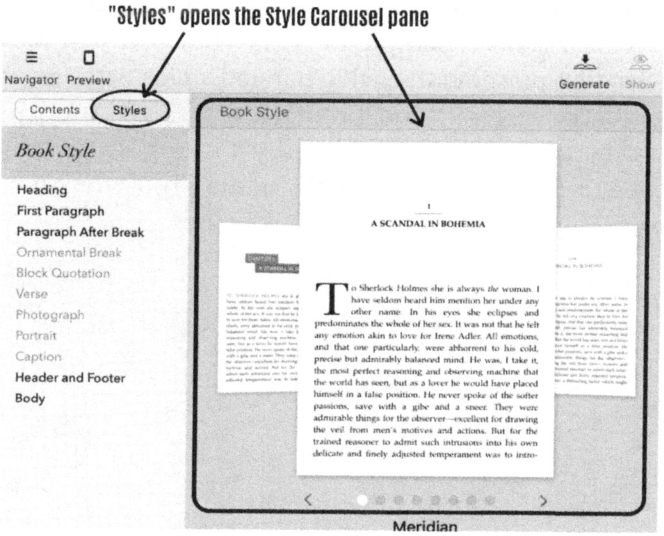

Fig. 4: Styles button and Style Carousel pane

Change Device. To see different e-reader device previews as well as the print view, use the Change Device option. Click the small tablet/phone icon at the top of the Preview pane (*Fig. 5*).

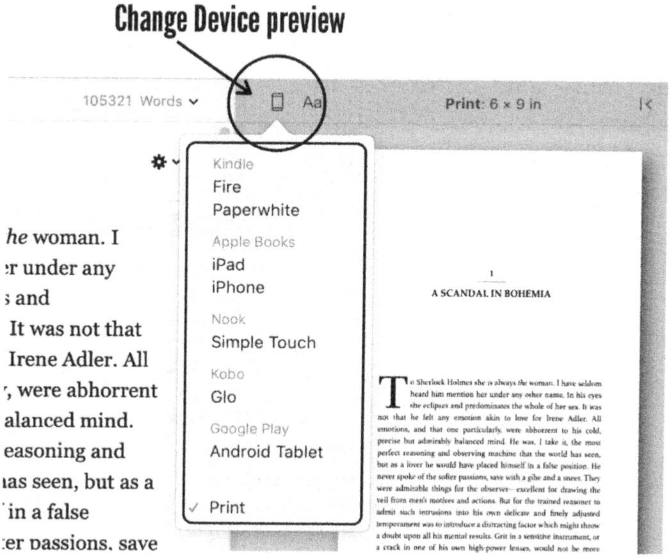

Fig. 5: Change Device preview options

If you choose Print view, the default appearance is a plain white rectangle with a generic sample of what your text will look like. To view precise text placement, click Change Device Settings where you can switch between Draft and Proof view (*Fig. 6*). You can tell at a glance which print view you are in by looking at the page edges; Draft view is a plain white rectangle, while Proof view has a faintly shaded page edge representing the book spine.

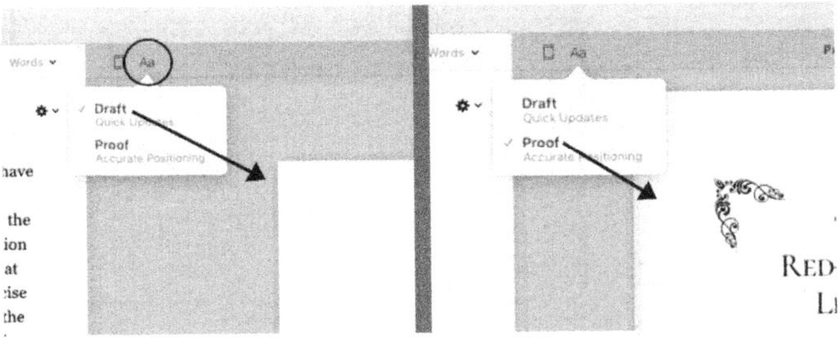

Fig. 6: Draft vs. Proof in Print view

Proof view could make Vellum run more slowly, so Draft view for print is shown by default.

TWO COMMON MENUS

Two of the most-used menus are selected by the "gear" buttons at the bottom of the Navigator (*Fig. 7*) and near the title text box (*Fig. 8*). I'll refer to these throughout the book.

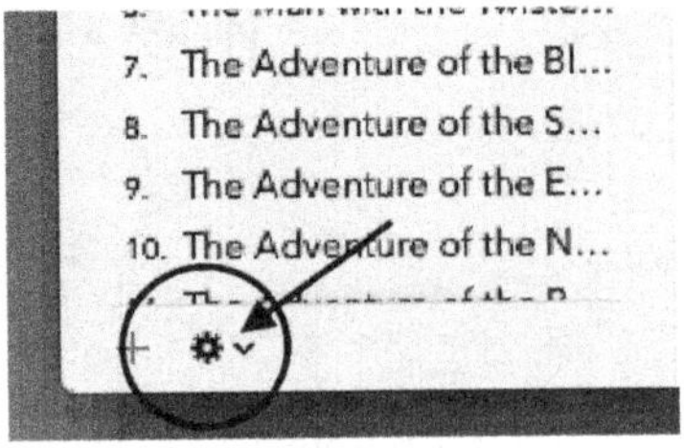

Fig. 7: Navigator gear menu

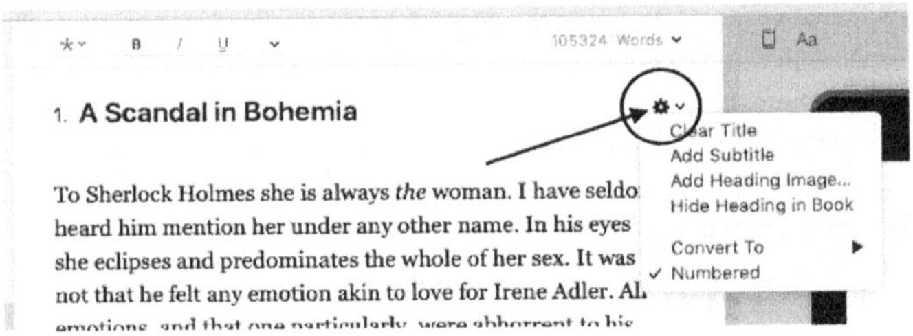

Fig. 8: Title box gear menu

NOTES:

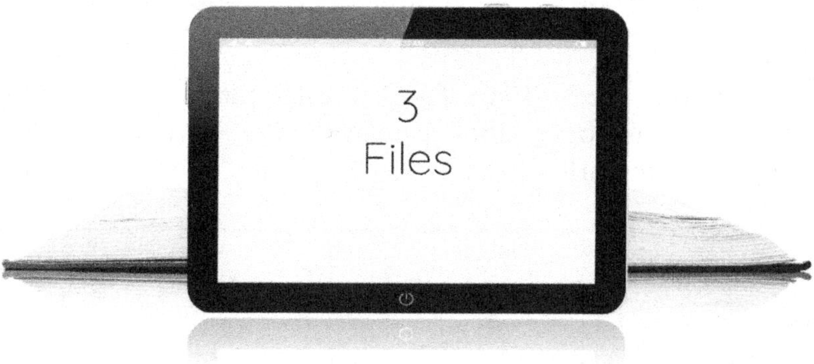

Let me begin by clearing up a couple of common questions. After importing a document, you will choose a working filename for each project which Vellum will access on your computer, server, or cloud storage. Vellum itself does not store these files, so I'll demonstrate how to save each project where you can easily identify and find it, whether you name your project "Experiment X-13," "Fred," or something boring-but-practical like "Sherlock Holmes - print."

In the "Setup" chapter, I'll show you where to input your book title, and that is where you need the wording to be exact.

Also, please visit the "New to Mac?" chapter if you need clarification of any Mac terms (like *Dock* or *Launchpad*).

BEFORE YOU BEGIN

Make sure your manuscript file is saved as a .docx extension. When you import your .docx file to Vellum, the import process should keep any bold and italic text overrides and detect chapter divisions, but will not keep specific fonts.

Fig. 9: Import Word File

3 WAYS TO IMPORT A .DOCX FILE

Option 1: Double-click the Vellum icon in the Dock or the Launchpad.

Click the Import Word File button at the bottom of the pop-up window (*Fig. 9*).

Open your .docx file location (*Fig. 10*). Click the filename and select Import. If a file is not in .docx format, then it will be grayed out as unavailable for import.

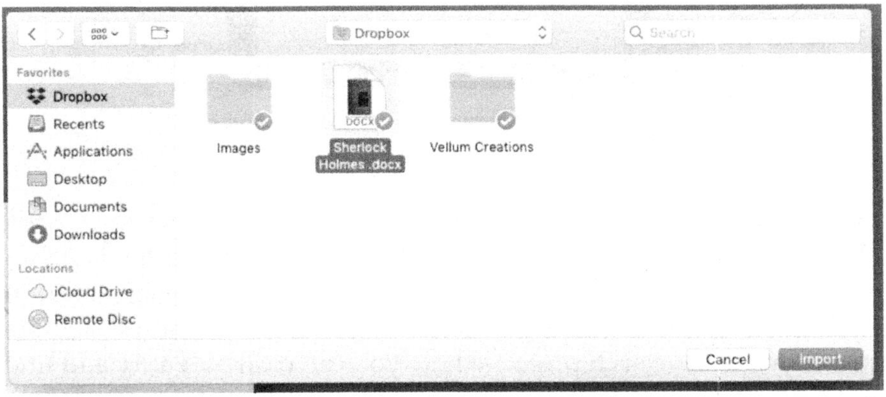

Fig. 10: Choose file to import

Option 2: Open the Finder window from the Dock. Navigate to your document's file location—no need to open Vellum first. Click and drag the file and hover it over the Vellum icon in the Dock (*Fig. 11*). When the program name "Vellum" pops up, release the mouse button, and the file will import as a new project automatically.

Fig. 11: Drag file to the Vellum icon

Option 3: If Vellum is already running, go to the Menu Bar, click File > Import Word File, and choose the desired file. This will open a new Vellum project.

PC USERS: A NOTE ABOUT SAVING FILES

Choosing Desktop as the location for your working files makes it easy to find what you're working on. However, for files that need to be accessed from your PC and from your second computer or Macin-

Cloud, a cloud file transfer system makes things simple. I personally use Dropbox, as their free storage has been plenty for my file access needs, so I've included a chapter just on connecting to Dropbox.

SAVING YOUR WORKING BOOK FILE

At the top of the Vellum window, you'll notice the default filename after import is "Untitled" (*Fig. 12*). Click on "Untitled" to bring up the screen where you can change the file name and specify where to save it. (Again, this does not have to be your book title, just the local working filename so you can find it.) Hit Enter/Return to save.

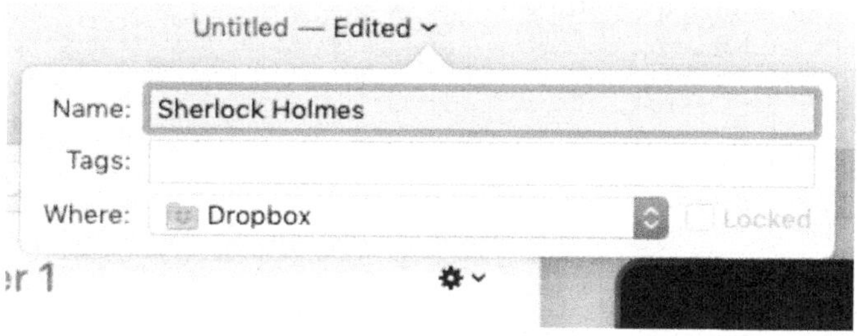

Fig. 12: Click filename to rename; hit Return to save

You can also update this via the Menu Bar by choosing File > Save.

The top of the window will show "[filename] — Edited" (*Fig. 12*) if the most recent changes are not saved. Vellum does have an auto-save feature, but it's a good idea to purposely save every so often.

WHAT ABOUT SEPARATE PRINT AND EBOOK FILES?

In some cases, creating two separate files for print and ebook versions is helpful. When I fine-tuned the print formatting of this book, for instance, I made some photo placements that were not logical for ebook, so another file was necessary. And for ebooks in general, I recommend left-aligned text, while print books are expected to be justified text—so having separate project files for these formats makes sense.

Before you create separate files, it's smart to make sure you've already sent your file to your editor, proofreader, and/or beta readers

(I highly recommend all!). Otherwise, any further corrections are made to multiple files rather than just one.

Creating a second file. Inside your finished file, go to File > Duplicate. Rename the new file something easily identifiable, like including "print" or "ebook" in the filename. Adjust the file copy as desired. (See the "Generating" chapter for more information.)

If you have too many or too few chapters after import, see the "Elements" chapter. Visit help.vellum.pub/importing for tips on how to optimize your file for conversion or to troubleshoot common issues after import.

NOTES:

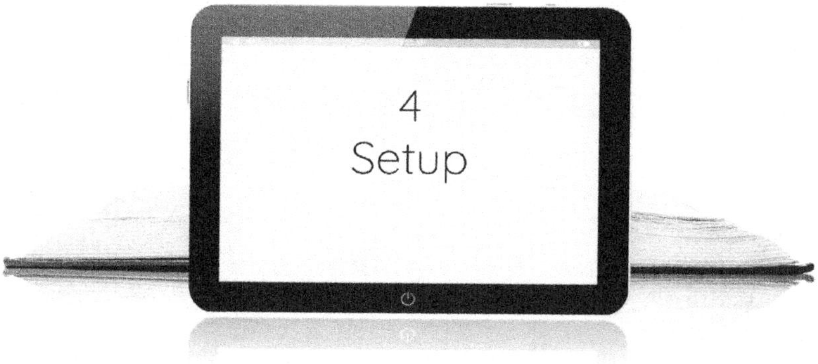

Most of the steps between here and the "Generating" chapter may be completed in no particular order. This particular chapter may not have the most exciting steps in your book format, but these steps are essential to properly set up your book file.

METADATA AND EBOOK COVER

The author and title information you enter sets up your title page. This information is also used in Vellum's print page headers and will be embedded into ebook files as the metadata used to identify and catalog an ebook on an e-reader device or app. (This information will also be used as the name of the folder where Vellum will generate files later.)

Fig. 13: My Book—title page information

To begin, make sure you are in the Contents menu and click on My Book at the top of the Navigator (*Fig. 13*).

Under the Title Info button at the top of the Text Editor pane (*Fig. 14*), type in the full book title, the subtitle if applicable, and the author's name.

Clicking the small "+" sign near the blank for Contributors (see *Fig. 14*) will allow you to add other project collaborators to the title page.

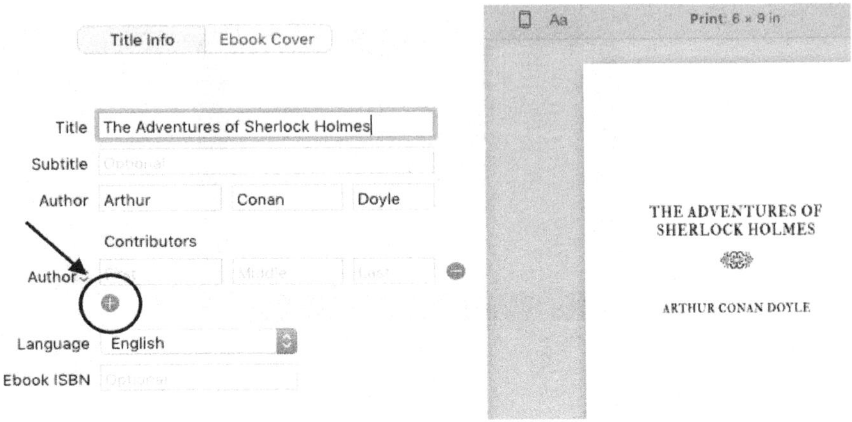

Fig. 14: Title Info—enter details for title page and metadata

Change the category by clicking on the default "Author" designation and choosing the appropriate option from the drop-down menu (*Fig. 15*).

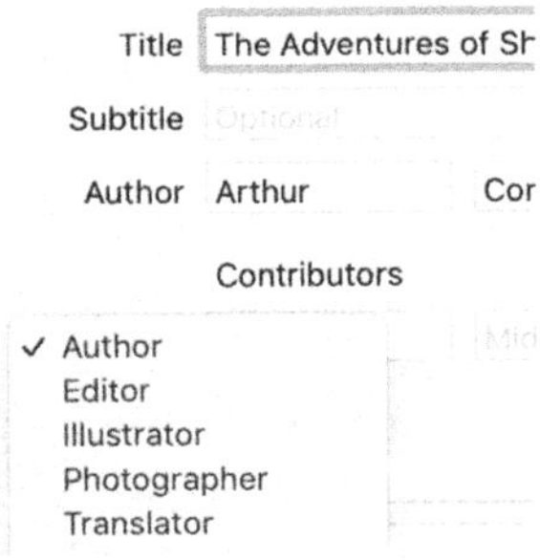

Fig. 15: Add other project contributors

The Ebook Cover button (next to the Title Info button) selects the screen where you can add a front-only cover image for ebook files. Under Ebook Cover, click the large "+" in the center of the image (*Fig. 16*). Choose your cover image file from the following screen and click Open.

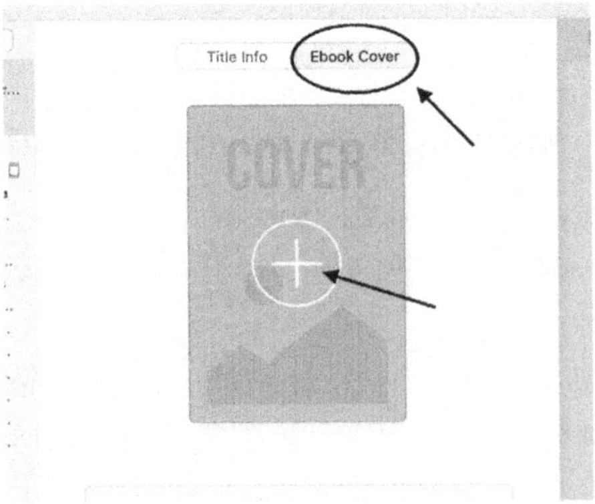

Fig. 16: Add cover image for ebook

TITLE PAGE

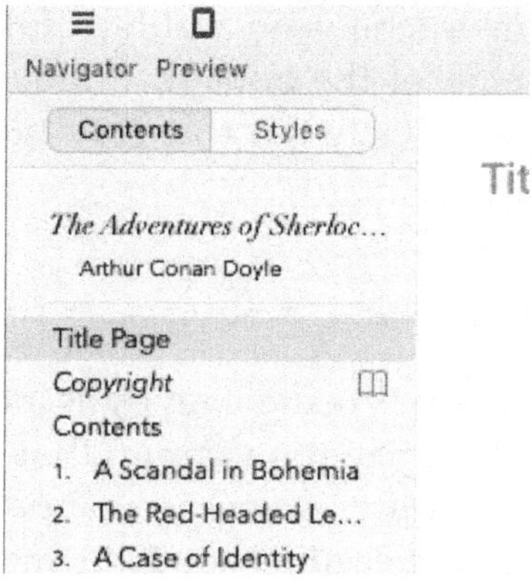

Fig. 17: Choose Title Page to add more publishing details

Vellum automatically generates a title page for each book file. Some of this information will be filled in from the My Book screen, but the Title Page selection in the Navigator (*Fig. 17*) is where you can click to open the Text Editor and add information like a publisher name, web address, or logo to the title page.

When you add a publisher web address in this window (*Fig. 18*), your ebook file will include a clickable link on the title page. The print file will not include a web link, so that information can be added on the copyright page in the next step.

Tip: For a publisher logo, the recommended image size is 281 x 314 pixels or larger. The image will always be small and centered at the bottom of the print title page.

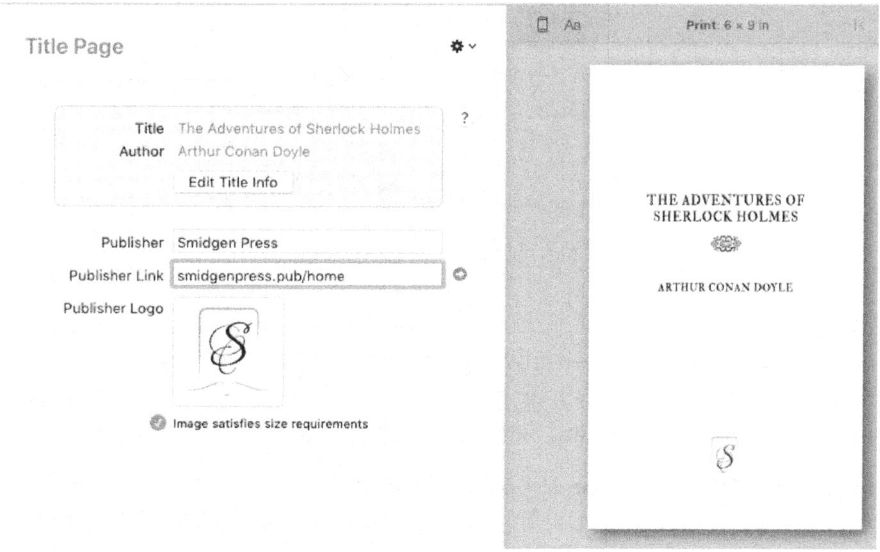

Fig. 18: Title Page editor

For print, if you wish to include a cover image in place of the text-only title page, see Adding Elements in the "Elements" chapter for instructions on adding a full-page image. Place this at the very beginning of the book (see Moving Elements in the "Elements" chapter), and delete the text-only title page. Your basic metadata will remain intact.

ADDING A COPYRIGHT PAGE

If you typed a copyright into your original document, Vellum should recognize it on import, but otherwise, a copyright is not added automatically. Add one manually by clicking the Navigator gear menu and selecting Add Element > Copyright (*Fig. 19*). Vellum will generate a basic copyright page based on the title and author you entered on the My Book screen.

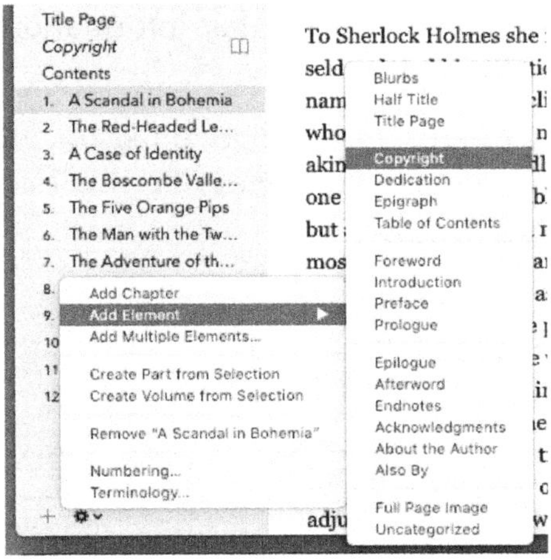

Fig. 19: Adding a copyright page

Use the Text Editor screen to customize the generic copyright by adding additional information like editor, illustrator, book designer, quoting permissions, ISBNs, etc. (*Fig. 20*). You also have the option to keep or omit the "Created with Vellum" notation.

Different copyrights for ebook and paperback may be appropriate if you are listing separate ISBNs or other version-specific information. In this case, select Add Element > Copyright twice in the same document to generate two copyright pages. Click the title box gear menu

(*Fig. 21*). Select Include In and choose from All Editions, Ebooks Only, or Print Only, and modify as desired.

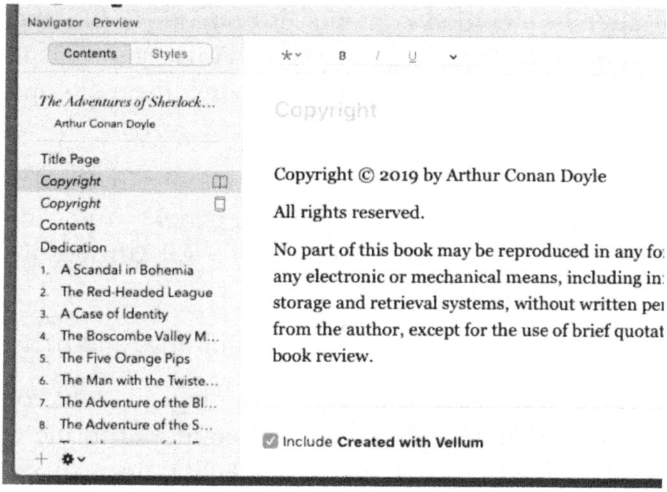

Fig. 20: Editing your copyright page

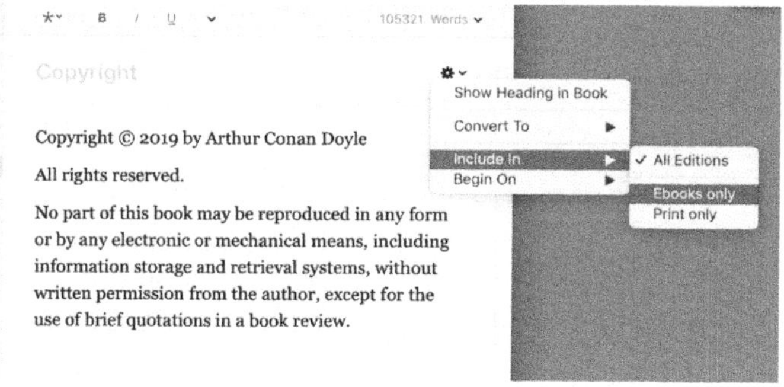

Fig. 21: Choosing which file(s) should include this element

If you choose only ebook or print, that element will show in italics in the Navigator menu (*Fig. 22*). To the right of the text, you will see either a phone/tablet icon for ebook or a book icon for print, specifying where it will be visible. This can be used for elements like Contents, as well.

Fig. 22: Icons show when items are in print only or ebook only

TABLE OF CONTENTS

Vellum generates an automatic table of contents in every project. If you have already included contents in your .docx file, they are not needed, so you may remove them prior to import or delete the extra material from the Navigator. (See Removing Elements in the "Elements" chapter.)

Many fiction genres are omitting the contents in print books, so Vellum defaults to only including them in ebooks. You can select the title box gear menu and choose Include In > All Editions if you want the contents in print as well.

See the "Box Sets" chapter for contents in volumes and parts.

Tip: Most ebook sales platforms require a "logical" table of contents (or NCX file), so Vellum will always generate that in the coded file, even if you remove the Contents element. But since ebook contents are trickier than print books, it's a safe practice to keep the contents in your ebook so online sales platforms are happy. For more reading on the subject, see help.vellum.pub/elements/toc.

NOTES:

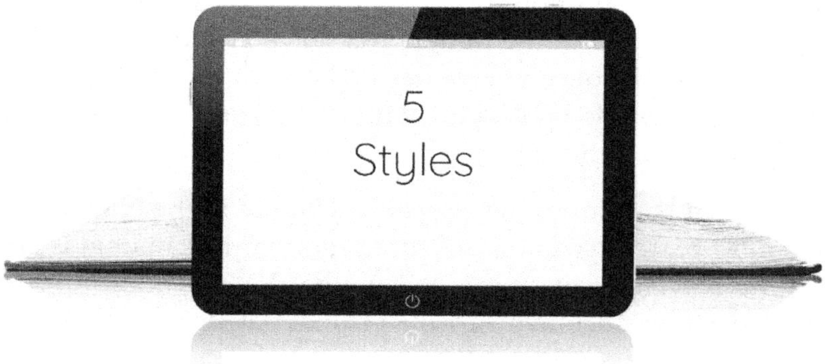

Adjusting your book style may be done at any point in the book formatting process. This step is enjoyable for some people but overwhelming for others. Feel free to skip this until later.

CHOOSING YOUR BOOK STYLE

At the top of the Navigator pane, switch from the Contents button to the Styles menu. When you click Book Style (*Fig. 23*), the Style Carousel pane will let you scroll through all book styles and view their chapter opener pages.

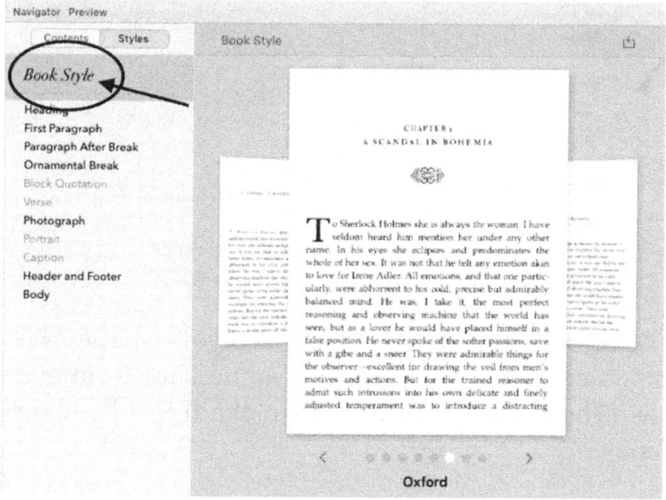

Fig. 23: Style Carousel

There is no "select" button for any of the Style Carousel options; the style simply updates to the option you leave at the forefront of this screen. (Be careful—once you've decided on a style you like, it's easy to change your whole book style without realizing it, just by clicking through the other options.)

I'll touch on a handful of the selections you see in the Navigator, but for the most part, the choices are straightforward. If an item is grayed out in the Navigator, it simply means that Vellum does not have it identified in your document. You can still define these settings.

HEADING

Each main book style has between three and six chapter heading options (*Fig. 24*). When you scroll through, you will notice that some styles have color heading options. While these will be the same text alignment in both print and ebook, the headings appear in color in ebook files but are only in black and white in the print PDF.

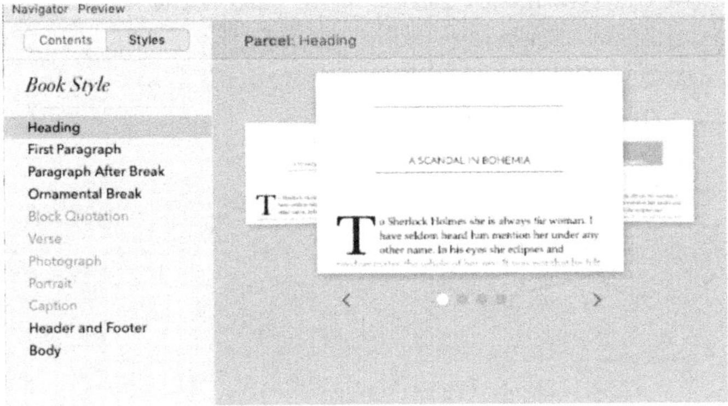

Fig. 24: Choosing a chapter heading style

Appendix A of this book contains all twenty-six chapter headings for you—that's a collection of all print heading choices in every Vellum book style. (The added color options make thirty heading possibilities for ebooks.)

FIRST PARAGRAPH

This option further customizes your chapter opener page. This is where you find drop caps, text in small caps, or the option for no custom text—just a regular paragraph.

ORNAMENTAL BREAK

Choose from the included section break images, or add your own by clicking Custom Image (*Fig. 25*). Custom images used as a break are very tiny on the page, so a simple image is best. Some break images vary by book style.

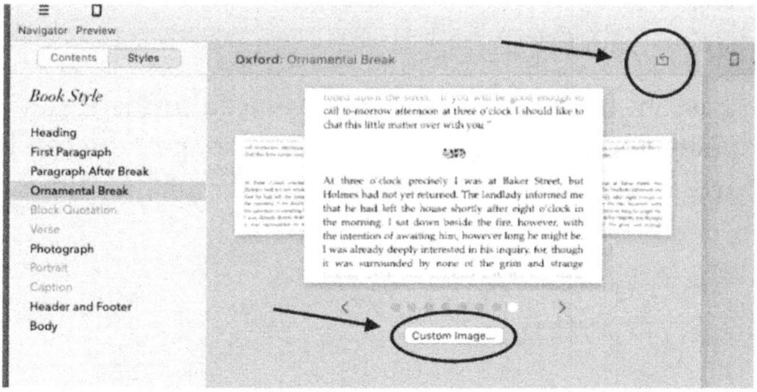

Fig. 25: Ornamental Breaks—use defaults or add a custom image; reset is at top right

To get back to the defaults after adding a custom image, click the square-ish arrow at the top right of the Style Carousel pane (*Fig. 25*).

BLOCK QUOTATIONS AND VERSE FORMATTING

Quotations and Verse selections can each be treated three different ways, at the time of this writing (*Fig. 26*). If you have trouble with these not being formatted properly, note that block quotes will *not* be styled this way in elements categorized as Blurbs, Dedication, Epigraph, Also By, or Uncategorized. Verse paragraphs will be styled like this in all main text elements except Also By. See the "Text Features" chapter for more details.

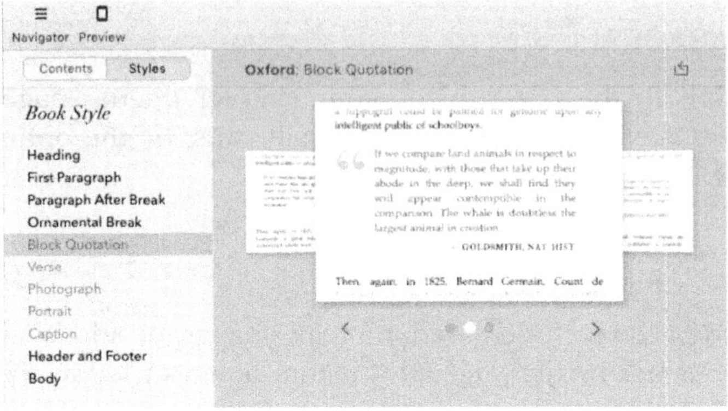

Fig. 26: Block Quotation options

PHOTOGRAPH AND PORTRAIT

These are custom treatments for images placed in the text. (The drop-shadow effects are not shown in print.) The image type can be chosen after import, as detailed in the Images section of the "Text Features" chapter.

HEADER AND FOOTER

There are several options for page headers and number placement (*Fig. 27*). These are used in print files only.

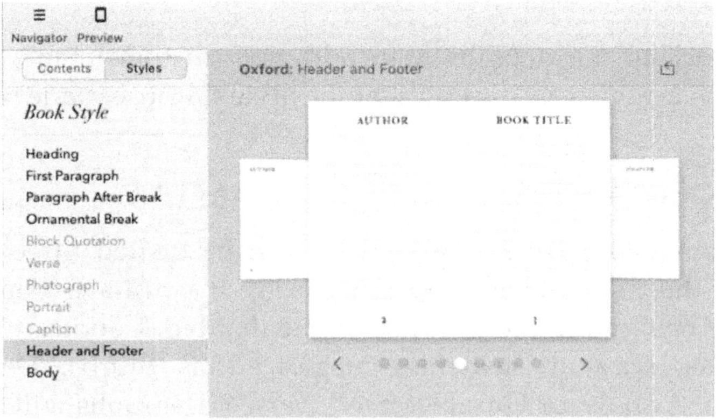

Fig. 27: Page header and footer options

Tip: A subtitle in the middle of a chapter that describes the section following it is called a head*ing*, while a book title or page number at the top of a page is called a head*er*. This can occasionally be confusing. Having items printed at the very bottom of a page, not surprisingly, is called the footer. Check your genre for what readers expect.

BODY

The Body selection (*Fig. 28*) has several adjustment options:

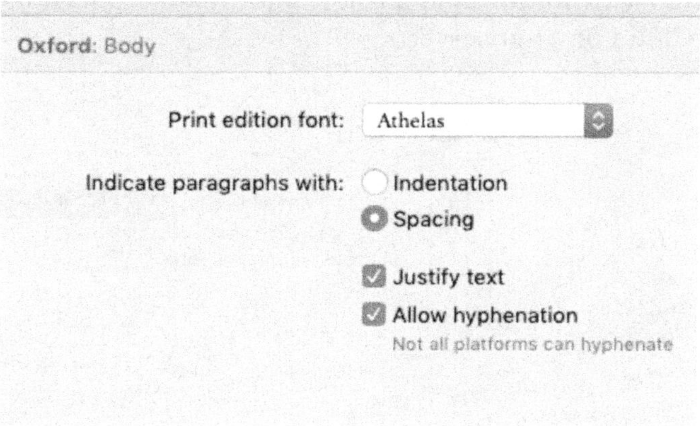

Fig. 28: Body selection options

- **Indicate Paragraphs**: To mark a new paragraph, choose Indentation (standard book style) or Spacing (web text style). Before choosing the spacing option, consider checking several highly-rated books in your genre so you know what your target readers expect.
- **Justify Text**: Choosing this means the book text will be straight along the left and right edges, which is standard for print books. Leaving this unchecked means the text will be left-aligned only, which is highly recommended for ebooks.
- **Allow Hyphenation**: If you don't care for hyphenated words, uncheck this option. Turning this off will likely cause awkward text gaps with justified text, so be aware.

Now—at the top of the Body options, let's look at fonts. (PC users, if you don't recognize most of these font names, it's because they're native Mac fonts that aren't common PC defaults.) You can't import a custom font into Vellum, but there is one additional font you can access as an Adobe subscriber or purchaser: Adobe Garamond Pro.

The first way to access Adobe Garamond Pro (AGP) in Vellum is to have it installed in your Font Book. The second is to have the Creative Cloud desktop app installed, be signed in with an active subscription, and have AGP activated in your Adobe Fonts. If all requirements are met, you'll see AGP as an option in the Print Edition Font drop-down (*Fig. 29*). To learn more about connecting Adobe fonts, visit help.adobe.com and search "fonts." (MacinCloud users: See the "MacinCloud" chapter for information on connecting to Creative Cloud on your server.)

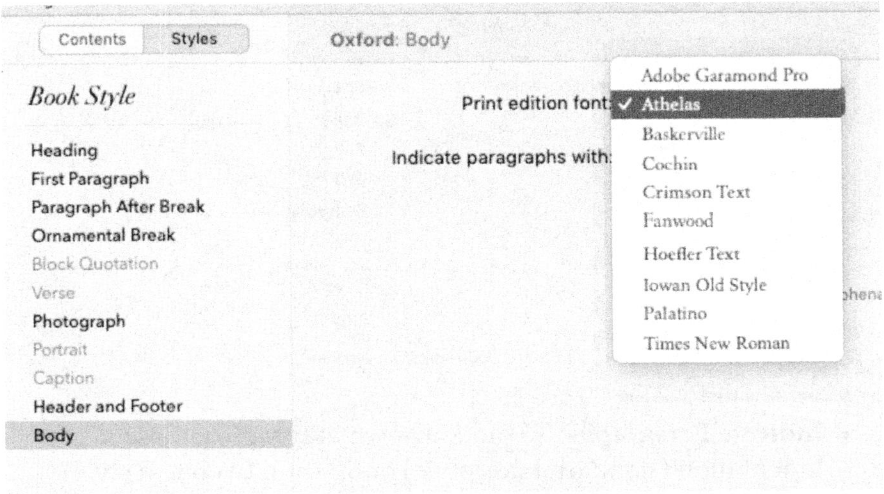

Fig. 29: Fonts for print PDF

To see what the various print fonts look like at actual size, turn to Appendix B for regular-size fonts and Appendix C for large print.

Tip: As indicated by the label "*Print* Edition Font," body fonts can only be specified for print books. Ebooks are always generated in a generic font, since e-reader settings can be customized by each individual user with their preferred font and text size.

NOTES:

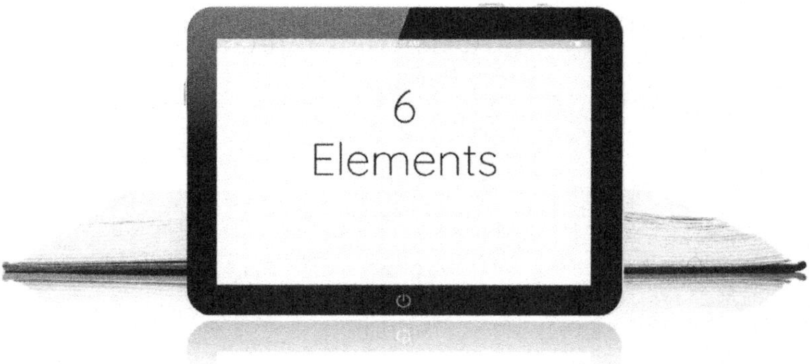

The basic elements you should include in a published book are title page, copyright, and main book text. Following are more details on element types, organization, and adjustment.

THE PARTS OF A BOOK

The elements of a book are categorized into these sections:

- front matter (items *before* the main book text)
- main body text
- back matter (everything that comes *after* the body)

On the next page, I've included a general overview of front matter, body, and back matter (*Fig. 30*). Of these, the half title is perhaps the least known; it's a smaller version of the title page with *only* the book title—no author or publisher name—on the first page of a print book.

The compiled list doesn't cover every book section, as the list of all possible items is quite long. For more information, I recommend "The Book Designer" website, especially Joel Friedlander's article called "Self-Publishing Basics: An Unabridged List of the Parts of a Book"—www.thebookdesigner.com/ 2009/09/parts-of-a-book.

Tip: You may notice that Acknowledgments are traditionally found in the front matter, but it's becoming more common in self-published books to move these to the back of the book to make best use of online "Look Inside" features. This way, readers get to the book's main content quickly.

TRADITIONAL PLACEMENT OF BOOK ELEMENTS

NONFICTION		FICTION
"Praise for" or half title page Title page Copyright Dedication Table of contents Foreword Preface Acknowledgments Introduction	FRONT MATTER	"Praise for" or half title page Title page Copyright Dedication Table of contents Acknowledgments Prologue
Main book text Afterword or Conclusion	BODY	Main book text Afterword or Epilogue
Endnotes Index/Bibiliography About the author Also by the author	BACK MATTER	Notes/Glossary About the author Also by the author

Fig. 30: Traditional order of book elements

REMOVING AN ELEMENT

As you fine-tune your manuscript elements and find yourself with ones that don't belong, use the Navigator pane to select the segment to be deleted.

From the Navigator gear menu, select Remove "[element name]" (*Fig. 31*).

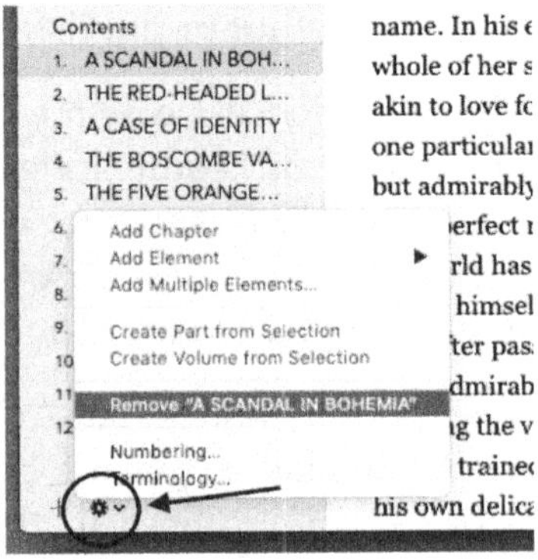

Fig. 31: Removing items

RENAMING AN ELEMENT

To rename, you have two options.

Option 1: Double-click the element name in the Navigator pane (*Fig. 32*) and change the description. This automatically updates the name here, in the Text Editor pane, and in the contents.

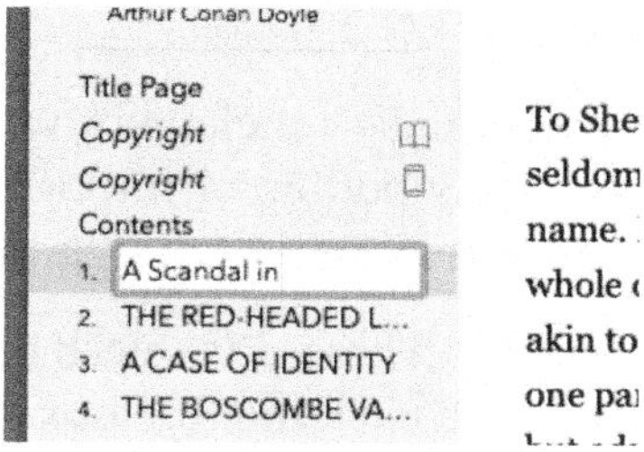

Fig. 32: Rename element by double-clicking in the Navigator

Option 2: Select the element in the Navigator, single-click the title box at the top of the Text Editor (*Fig. 33*), and change the information there.

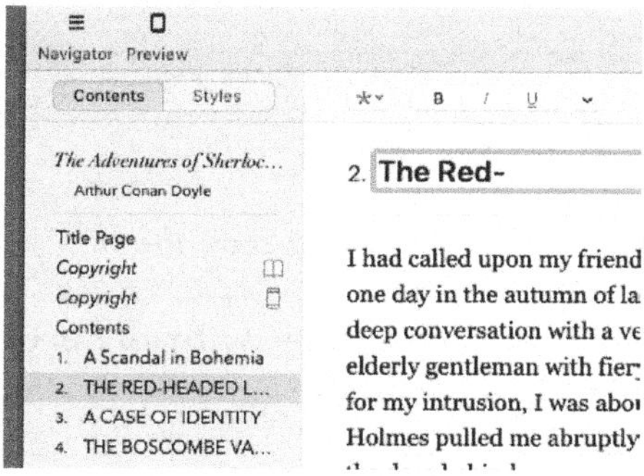

Fig. 33: Rename element at top of Text Editor

You can also add chapter subtitles here by clicking the title box gear menu and choosing Add Subtitle (*Fig. 34*).

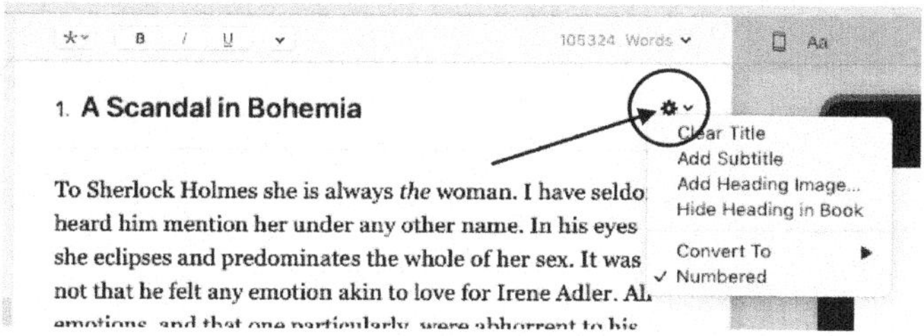

Fig. 34: Add chapter subtitle

RE-CATEGORIZING

Occasionally an element will be imported but will be mislabeled. For example, perhaps I imported a document with an unlabeled dedication which Vellum designated as Chapter 1.

To correct this, I'll select the mislabeled Chapter 1 in the Navigator pane. I'll click the title box gear menu, then choose Convert To > Dedication (*Fig. 35*).

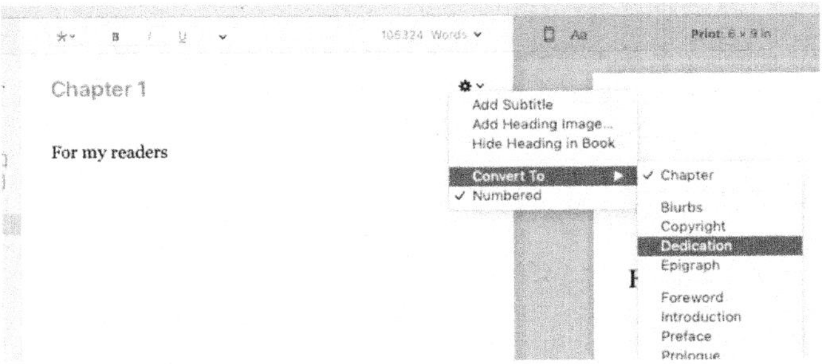

Fig. 35: Recategorizing an element

This process works for any element, with the exception of the Title Page and the Contents pages; these may be deleted but may not be changed to a different element.

MOVING ELEMENTS

The Navigator provides the simplest way to shift the order of items—just drag and drop items to rearrange them. For chapters, parts, or volumes that are automatically numbered, moving an item will renumber it sequentially.

ADDING ELEMENTS

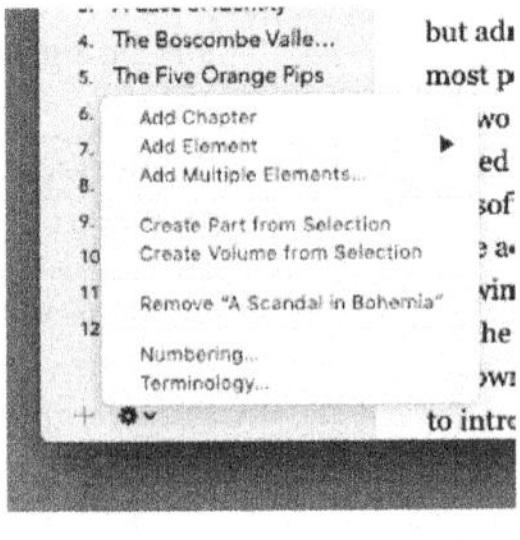

Fig. 36: Add Element

Adding one chapter quickly. Clicking the "+" is a quick way to add a blank chapter. Drag and drop to move the blank chapter as needed. Alternatively, choose Add Chapter from the Navigator gear menu (*Fig. 36*).

Adding one element of any type. To access

the list of all the single elements you can add, select the Navigator gear menu and choose Add Element (*Fig. 36*).

Let's add a full-page image as an example.

Click the Navigator gear menu and choose Add Element > Full Page Image (*Fig. 37*).

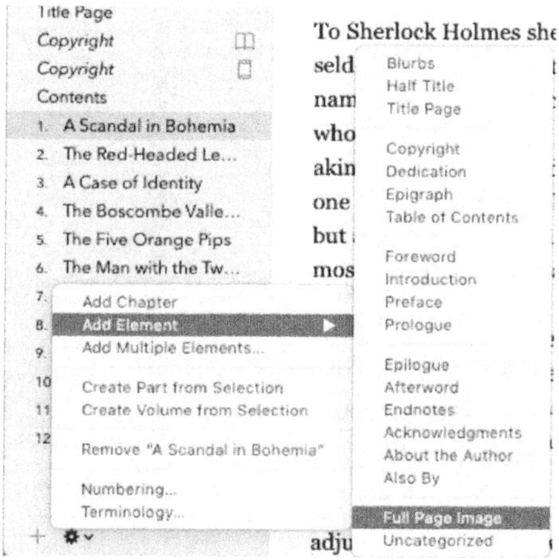

Fig. 37: Adding a full page image

Hover your mouse over the image blank that pops up, click the large "+" sign, and select the image to add. If your Vellum version is new enough to support images with bleed, choose Print Span—one page or two. Then select Print Extent—regular margins, safe zone, or full bleed. (The "safe zone" option is almost full-page but leaves 1/4" around the image to meet "no bleed" standards for Print on Demand sales platforms.)

Click and drag the Full Page Image marker up and down the Navigator to place it where you want it. This must go between other elements.

These images default to a right-hand page, but you can override this by going to the title box gear menu for this image and choosing Begin On > Either Side of Spread. Here, you can also use the Include In option to specify which file type(s) should have this image present.

Adding multiple elements. The Add Multiple Elements selection

(*Fig. 38*) is extremely efficient for adding multiple types of items into the manuscript.

Click the Navigator gear menu and choose Add Multiple Elements. Check the items you want to include, then click Add at the bottom of the pane—or choose Cancel to exit with no changes.

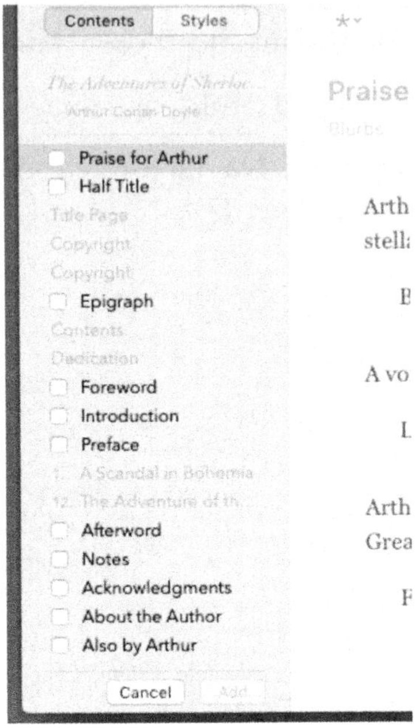

Fig. 38: Add Multiple Elements options

MERGING OR SPLITTING CHAPTERS

Occasionally, chapters need some fine-tuning. It happens—sometimes whether it's your first document import or your twenty-first.

Merging. A common issue in formatting nonfiction books is when Vellum mistakes subheadings for new chapters. When this happens, click through the sections and identify the ones that belong together.

In the Navigator, click on the sections to merge. Then hold Command + click to add one additional section at a time or Shift + click to select a range of items at once.

With all the items selected, go to the Menu Bar and click on Chapter > Merge Selected Chapters (*Fig. 39*); all highlighted items will be unified into a single chapter. The following chapters will automatically renumber.

Repeat, if needed, for any other sections.

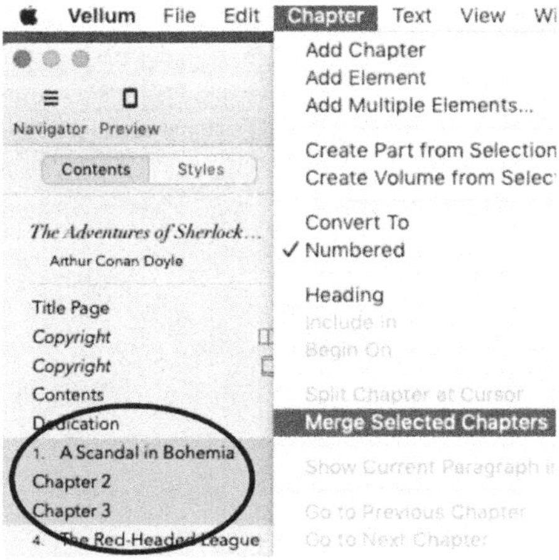

Fig. 39: Merging items into a single chapter

Splitting. The opposite issue arises when Vellum doesn't recognize *enough* sections. Here's how to manually split a chapter.

In the Navigator, select the chapter or section that needs to be divided. Don't change the beginning part of the section; just scroll down in the Text Editor to where the current chapter should end.

Position your text cursor at the beginning of the first word of the next chapter, then go to the Menu Bar and choose Chapter > Split Chapter at Cursor (*Fig. 40*). A new chapter will appear in the Navigator pane, where you can rename it as desired.

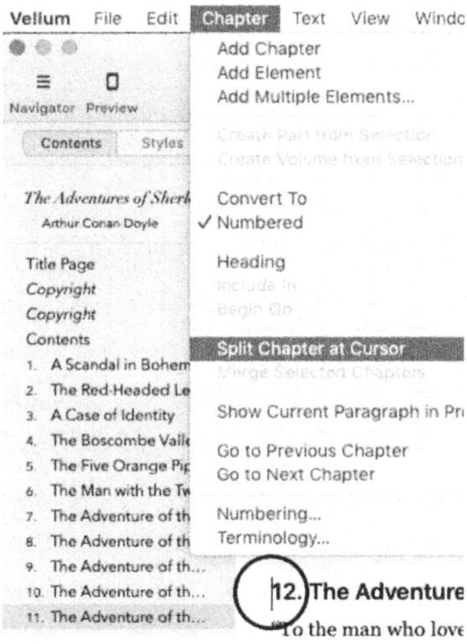

Fig. 40: Splitting a section into multiple chapters

NOTES:

Want to center or right-align a paragraph? What about adding a numbered or bulleted list or a decorative break? The Text Features drop-down (*Fig. 41*) is where you'll find that. Here is a summary of several options. Full features can be found online at help.vellum.pub/text-features.

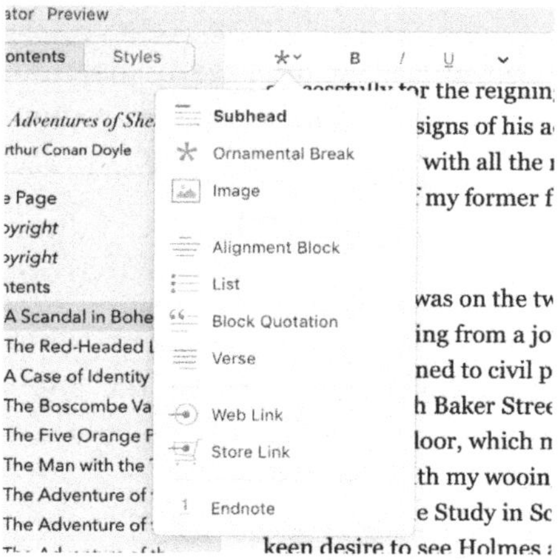

Fig. 41: Text Features menu options

SUBHEADS

To set text as a subheading, highlight the text you want, click the Text Features drop-down, and choose Subhead. To remove the Subhead formatting, click onto that line of text, click the gear menu to the right, and choose Clear Formatting.

There is only one Subhead setting in each book style. If you don't care for the appearance, you can either experiment with a different overall book style or manually adjust body text instead. Create your own subhead workaround using the bold or italic text overrides or by choosing Small Caps or Sans Serif from the Additional Formatting drop-down menu (*Fig. 42*).

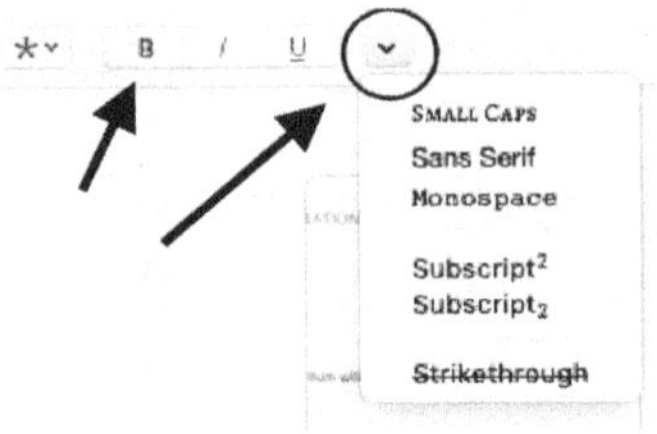

Fig. 42: Additional Formatting

ORNAMENTAL BREAKS

To place an ornamental break, position your cursor in a blank line of text between paragraphs. Choose the Text Features drop-down and select Ornamental Break. See the "Styles" chapter for more details.

IMAGES

Images will be imported if they are already inserted in your .docx file, but using the Text Features menu is one way you can insert images directly. File types supported are JPEG, PNG, or TIFF.

To insert, bring up the section in the Text Editor pane where you want to add the image. From here, there are two options:

Option 1: Photo Library. For photos on your computer or connected in your iCloud photos, you can go to the Menu Bar, then click Window > Show Photo Library (*Fig. 43*).

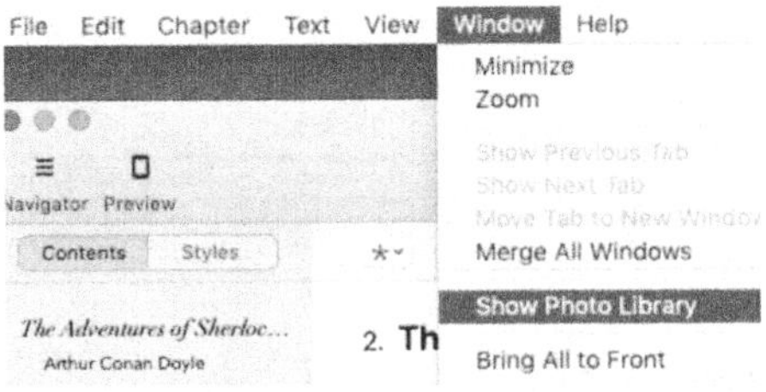

Fig. 43: Show Photo Library

Drag and drop the photo where you want to place it. Drag carefully

—it will split a paragraph if you drag to the middle of a paragraph and not the end!

Option 2: "Image" from drop-down menu. If your desired images are located in Downloads, Desktop, or any location other than the Photos folder, place your cursor in between the paragraphs where you want the photo to go. Click the Text Features drop-down and choose Image. Select file and click Choose.

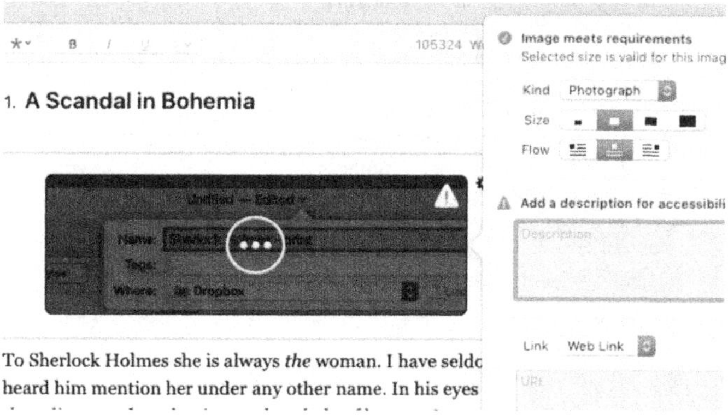

Fig. 44: Open the image settings

Clicking the center of the image will show three dots (*Fig. 44*) where you can click again to access the settings panel for that image. Choose the kind of image (default is Photograph), the size (based on the width of the image—not the height), and, for the two smallest sizes, the flow: left, center, or right. Use the Change Preview options to see how the image will appear on various devices.

The gear menu to the right of the image (*Fig. 45*) will bring up the options to add a caption or replace the image.

For the full range of details on managing inline image options, visit the tutorial at help.vellum.pub/text-features/inline-image.

Fig. 45: Image gear menu options

ALIGNMENT BLOCK

To change a paragraph's alignment (left, center, or right), highlight the desired text, open the Text Features menu, and choose Alignment Block. Text will be centered by default.

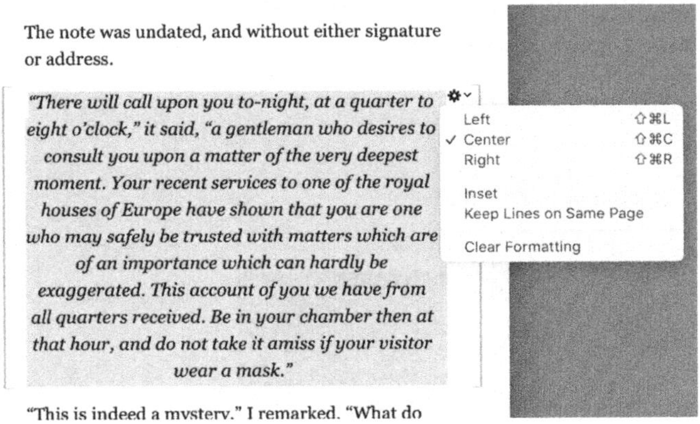

Fig. 46: Alignment Block formatting

When you click inside the aligned text segment (*Fig. 46*), the gear menu drop-down on the right allows you to specify several things: (1) how the text should be aligned, (2) if there should be an additional left indent ("Inset"), and (3) if the block should have the text kept all together (in the case of a poem or quote, etc.). The Clear Formatting option returns the selection in line with the preceding text.

LIST

To change text to a bulleted or numbered list, highlight the text and choose List from the drop-down menu. Clicking on the section's gear menu (*Fig. 47*) will let you select bullets or numbers.

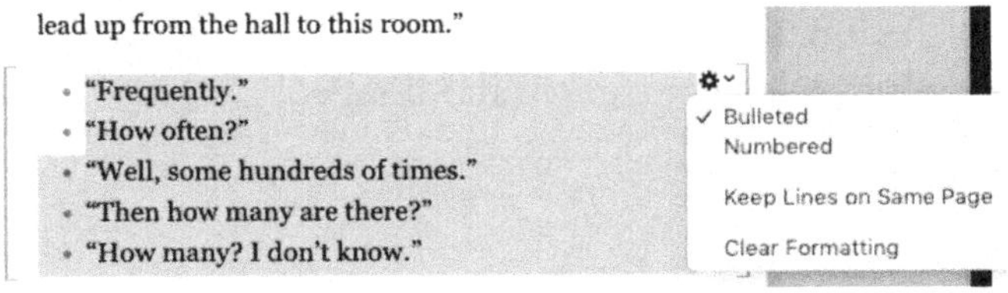

Fig. 47: List—bullets and numbers

BLOCK QUOTATION AND VERSE OPTIONS

For a quote, pullout box, poem, etc., highlight the desired paragraph and choose Block Quotation or Verse. Each of these selections has three style options in the Styles Carousel. Block quotes are left-aligned while Verse styles all have hanging indents (*Fig. 48*). See the "Styles" chapter for customization options.

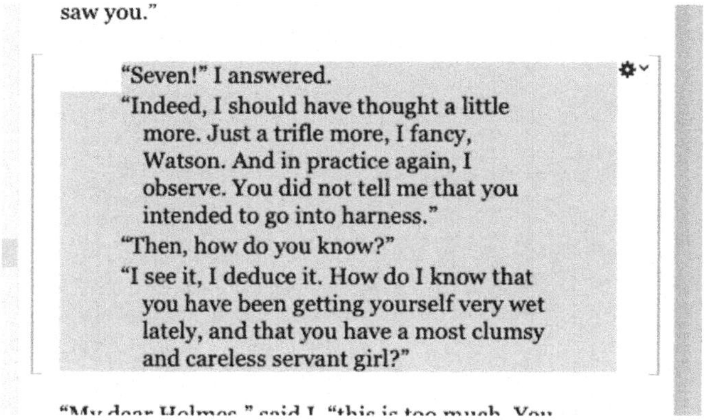

Fig. 48: Verse options have hanging indents for long lines

ENDNOTES

If you have formatted footnotes or endnotes in your .docx file, Vellum will import them, but only as endnotes.

If you want one main Notes section at the end of the book, but notes are appearing only at the ends of chapters, go to the Navigator gear menu and select Add Element > Endnotes. They should all appear at the end of the project. (Troubleshooting: Sometimes this step is necessary to create linked notes, even if a section called Notes was identified at import, depending on how the original document was formatted.)

Conversely, if you wish the notes at the end of the chapters and they are all at the end, delete the standalone Notes element and the endnotes will only be listed at the end of each chapter.

NOTES:

Custom touches at the beginning of a chapter can be a great way to connect with your readers. There are several ways to take advantage of this customization feature in Vellum.

ADDING ONE SMALL CHAPTER HEADING IMAGE

At the start of the chapter, click the title box gear menu and choose Add Heading Image (*Fig. 49*).

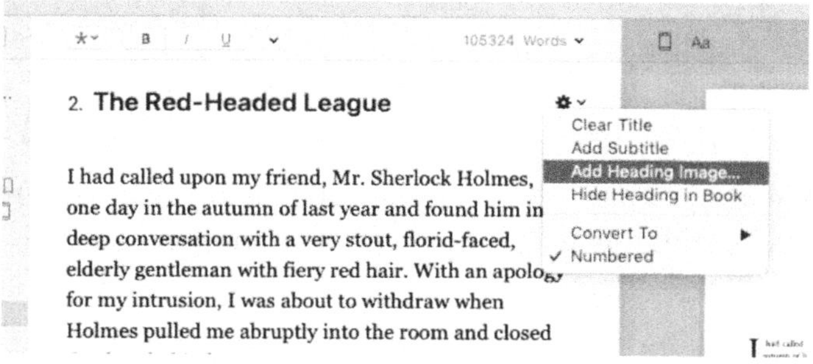

Fig. 49: Adding a small image or custom flourish for one chapter

Select the desired image, flourish, or other small item, and then click Choose.

This will add the image to the current chapter only. There are slight variations in size and image placement based on the book style and chapter heading you have selected.

ADDING ONE IMAGE FOR ALL CHAPTERS

Select multiple sections in the Navigator (with Shift + click or Command + click) where you wish to apply the heading image. With them all highlighted, choose Chapter > Add Heading Image from the Menu Bar (*Fig. 50*), then select the desired image, flourish, etc., and click Choose. This will apply that single small image to all chapters, keeping your ebook file size to a minimum, as this only counts as one "image" in the file.

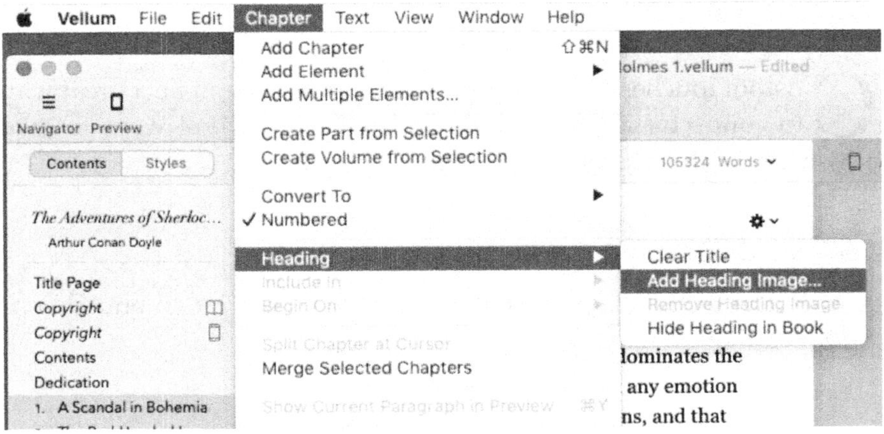

Fig. 50: Adding one small image to all chapter headings

If Add Heading Image is not an available option, you may have selected an item that is not designated as a Chapter.

USING A CUSTOM IMAGE AND REMOVING DEFAULT TEXT

Using a large custom image completely in place of the default chapter title text is one way to make your chapters memorable.

Custom heading images frequently include the chapter number or title along with decorative elements. To be a high enough resolution to stay sharp at the widest setting, I have found that an image a minimum of 5 inches wide (480 pixels) or wider works best for print. The custom heading images for this book were created at 5 inches wide and 2.25 inches high and are displayed at full width.

Keep in mind that the image should also be clear enough to read on a phone screen—so very intricate designs may be unsuited for an ebook file. Check the different device preview options to compare readability, and don't be afraid to duplicate the Vellum file completely and set separate options for ebook and print (see the "Files" chapter).

To add a fully custom image heading, open the desired chapter location, click the title box gear menu, and choose Hide Heading in Book (*Fig. 51*).

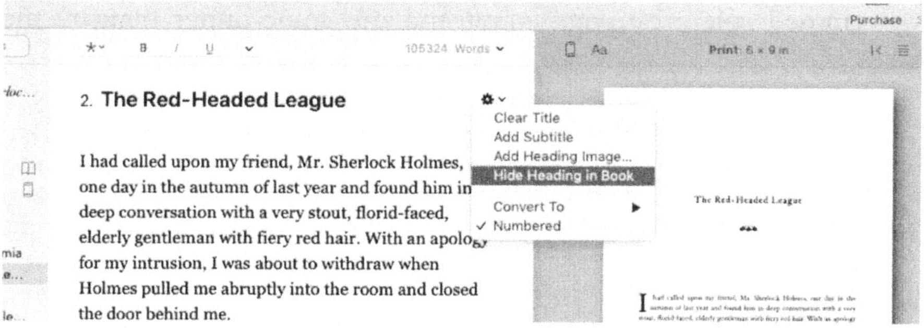

Fig. 51: Hide the text heading in this chapter

Insert the custom image at the start of the chapter text. Click the center of the image to choose the size setting and type in the image description (*Fig. 52*).

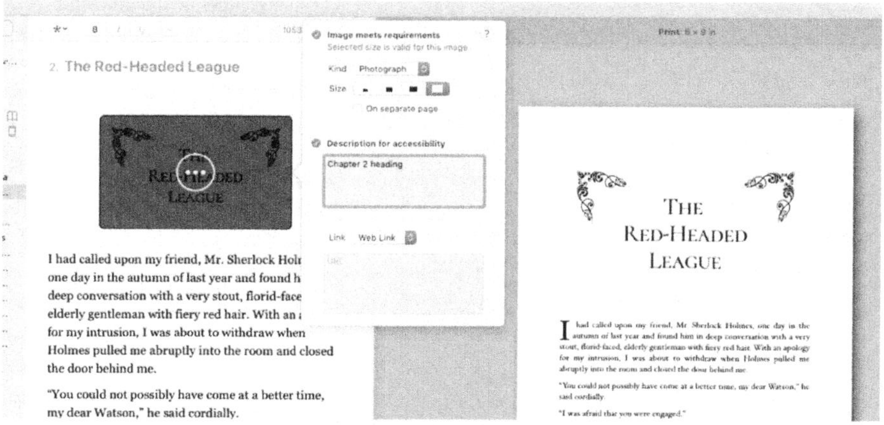

Fig. 52: Replace text heading with image

The default image designation is Photograph, as seen above the size selector. If you have styled the Photograph options in the Style Carousel, you can set this image to Freestanding instead, which will not add any extra borders or shading.

Tip: Using multiple images, especially large ones, may result in a higher "delivery fee" for ebook sales than if you use a single small image for all the chapters. This could be a difference of a few cents or a couple of dollars, depending on the size and number of images. (The ebook version of this volume, for example, has over 120 images, and the delivery fee is $1.34.) The cost of a print book, however, is determined solely by page count and ink color rather than by file sizes.

NOTES:

The following selections apply to print PDF formatting only.

TRIM SIZE

The width and height of a print book is called its *trim size*. Some common trim sizes are 5″x8″ and 6″x9″, but there are several options to choose from in the Print Settings menu.

From the Menu Bar, click File > Print Settings (*Fig. 53*).

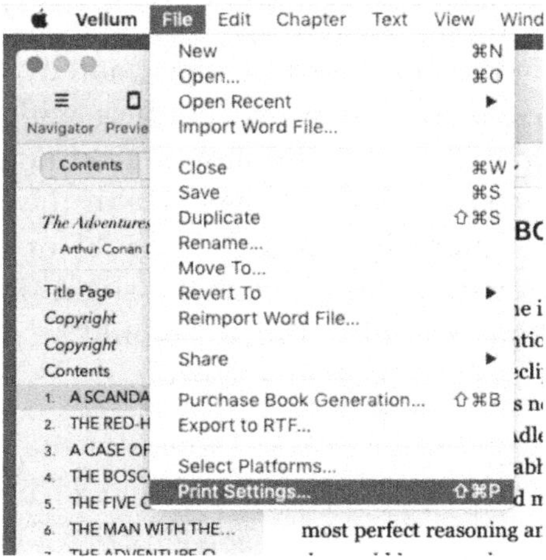

Fig. 53: Print Settings menu

In the Print Settings dialog box, you can quickly choose from four of the most common trim sizes (*Fig. 54*). Click the More Options button below those top four selections to see all available trim sizes, including the three book sizes in which Vellum offers large print text. (Large Print is a print-only setting, since e-reader size settings, like the preferred screen font, are set on each individual device or app.)

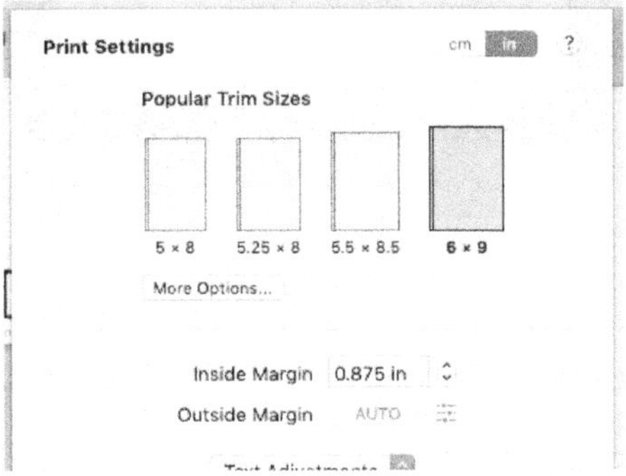

Fig. 54: Trim Size options (with More Options button)

FULL BLEED SETTINGS AND TRIM SIZE

If you use the Full Page Image option, add an image, and choose Full Bleed, this will not change the overall trim size of your book. It will potentially affect the size of image creation and will change the settings you choose when you send your book to print.

A quick explanation of full bleed: if you wanted your print book to have a solid black page—no white border—with white text on it (which is actually the paper color showing through, not white ink), then you would need to set the image size of the black rectangle to extend beyond the outer edges of the page 1/8″ or .125″. If you made the black image exactly the page size, the slight deviations as paper shifts in the machine trimming process would be obvious. Allowing for bleed ensures true full-page image printing.

So creating an image for a 6x9 book should actually measure 6.125″ wide (no bleed needed at the spine edge) and 9.25″ tall, with bleed at top and bottom. Plan the image—especially if it's meant to be a two-page spread—so it does not include important items in the center where they might get lost in the binding.

Wherever you upload your book for sale, with the normal process of selecting your trim size (don't factor in the extra bleed measurements), the system should ask whether your file has bleed or no bleed. If you have even one full-page image with bleed, Vellum will automatically configure your entire print book to match document requirements. For full details and more examples from 180g, visit help.vellum.pub/elements/full-page-image.

Bleed settings do not affect ebook image display.

MARGINS

As seen in *Fig. 54,* the Print Settings window shows the default margin settings below the trim sizes, and you have the option to make custom adjustments. (*Tip*: Nonfiction books frequently have more generous margins than fiction books.)

TEXT ADJUSTMENTS AND OTHER ITEMS

These fine-tuning options are also found in the drop-down menu at the bottom of the Print Settings dialog: Text Adjustments, Chapter Begin, Page Numbering, and Images (*Fig. 55*).

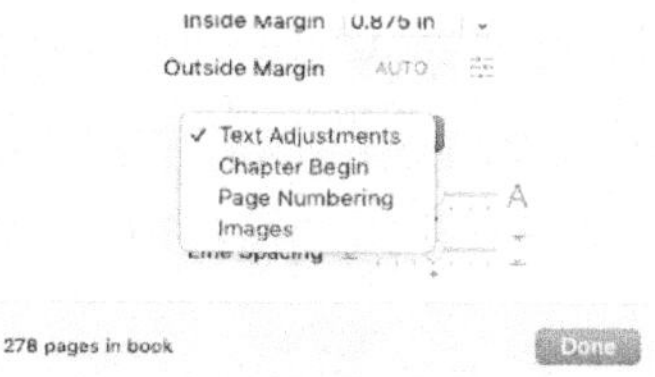

Fig. 55: Drop-down menu—click on Text Adjustments

Text Adjustments. The sliders under Text Adjustments (*Fig. 56*) allow you to fine-tune your print file's text size and spacing. The top slider allows you to increase or decrease the font size, and the bottom one adjusts spacing between lines. If you have chosen a Large Print trim size, these settings will automatically be for large print font sizes.

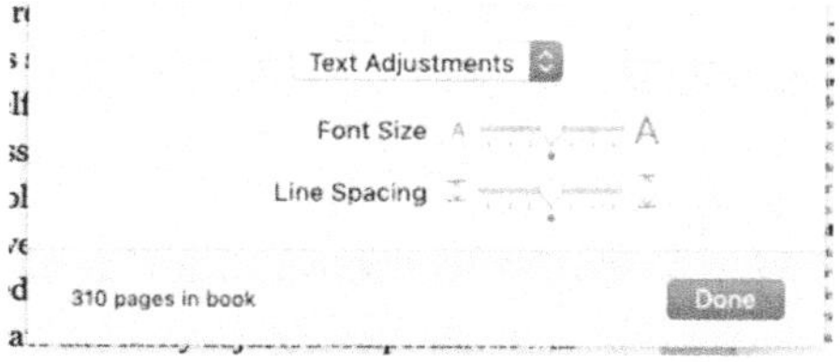

Fig. 56: Text Adjustments sliders for font size and line spacing

Chapter Begin. Clicking on Text Adjustments will reveal more options. The second setting is Chapter Begin (*Fig. 57*), where you can override the default setting of all chapters beginning on a right-hand page.

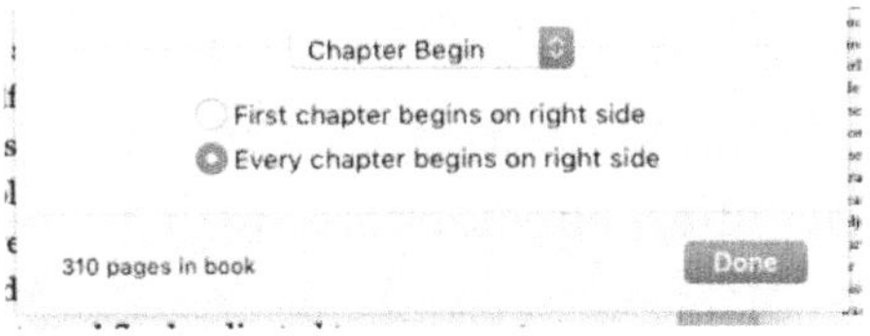

Fig. 57: Chapter opening page options

Tip: Many nonfiction print books put out by traditional publishers begin every chapter on a right-hand page, even if that leaves a blank left-hand page opposite. Fiction works commonly have Chapter 1 begin on a right-hand page but let the rest of the chapters begin on the next available page to maintain the flow of the story.

Page Numbering. The next drop-down is where you can adjust the Page Numbering option (*Fig. 58*).

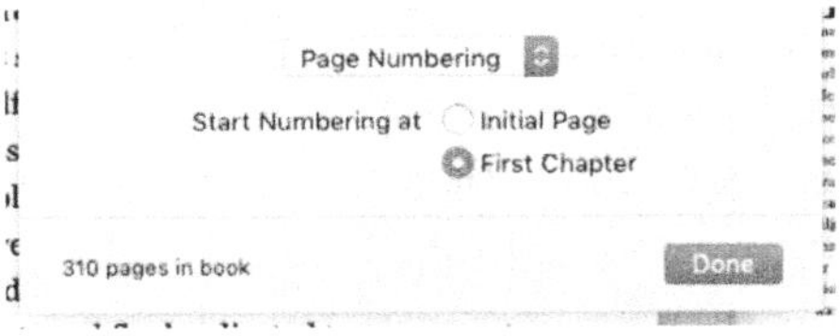

Fig. 58: Adjust page numbering

Most books begin with page 1 on the first page of Chapter 1, with lowercase Roman numerals marking the preceding pages. If this is a book for publication and not a small booklet for general use, there is generally no need to change this.

Images. The final Images menu simply specifies whether print images should be treated as color or as black and white. This is the print-only setting, as ebook images will show on the reader's exactly as they are imported, according to the e-reader device capability.

Tip: Even if you set images to export in color, you can still choose black and white printing to save on printing costs and maximize profits.

AND SORRY . . . CUSTOM LINE SPACING IS LIMITED

Adjusting your file for print can be a little different than you expect. If you're used to adjusting page layout in a word processing program by hitting the Return/Enter key multiple times, you'll find that doesn't work in Vellum. (To satisfy ebook format requirements, only one blank line of text is allowed by default.)

There are some "cheats" out there for print files, but ***for a file meant for ebooks, please don't try to manipulate the text spacing***. That's a great way to risk unhappy readers and poor reviews—and possibly to get your book pulled from distribution.

FINAL CHECKS

If you want to know precisely what your print file will look like before you generate the PDF, use the Change Device Settings button (*Fig. 59*) to view your project in Proof view instead of the default Draft mode. A faintly shaded edge indicates the book spine.

Since Vellum can run a bit slower in Proof mode, Draft is perfectly fine for general work.

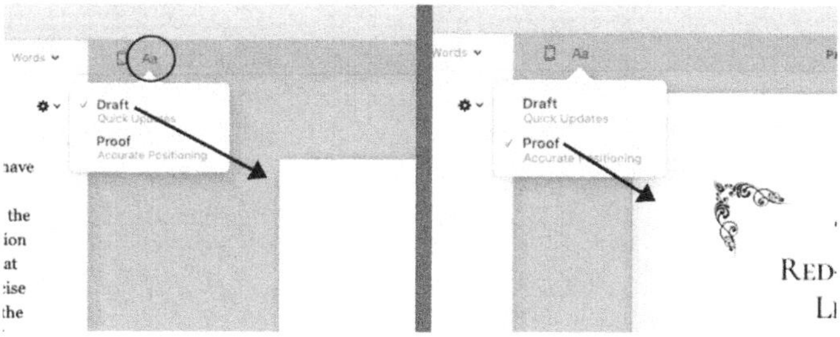

Fig. 59: Draft vs. Proof view for print

NOTES:

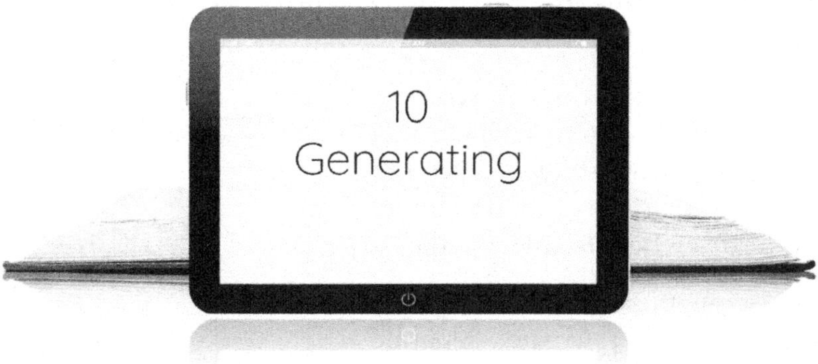

10 Generating

If you've followed the guidelines listed in this book, you should have your basic book setup DONE! You can use every Vellum feature in the program for free until this step: file generation.

GENERATING BOOK FILES

At the top right of the Vellum window there is an icon labeled Generate (*Fig. 60*). If it still says Purchase, you can click through and purchase a license or restore a previous purchase.

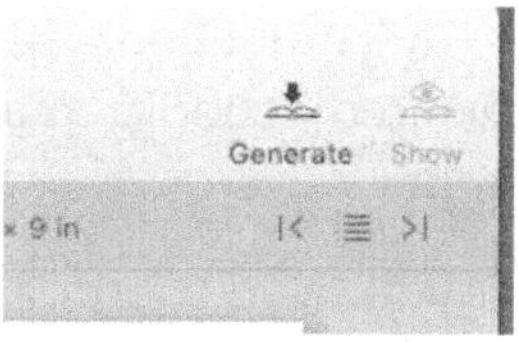

Fig. 60: Generate icon

You need the free Kindle Previewer software to generate Kindle files. This is already installed on MacinCloud along with Vellum, and users who work on a Mac will be prompted with a direct link to download Previewer if they haven't already done so.

Click Generate to open the Generate Books window or go to File > Generate Books from the Menu Bar (*Fig. 61*).

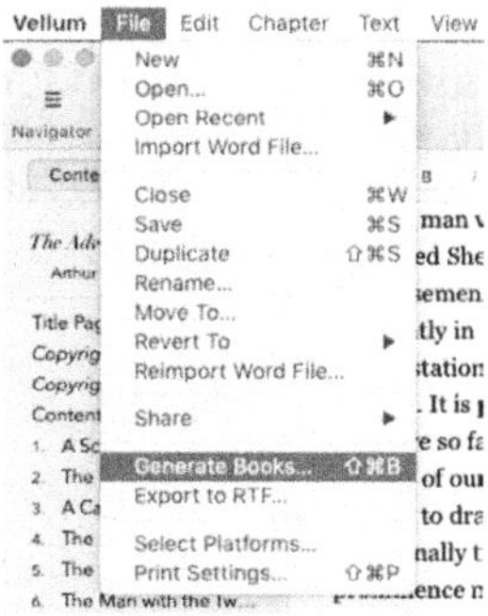

Fig. 61: File > Generate Books

At the right of the popup window, you'll see two small circles (*Fig. 62*).

The top circle with three small dots will open up the window where you can choose which file types to generate (*Fig. 63*), while the lower one indicates where to save the files.

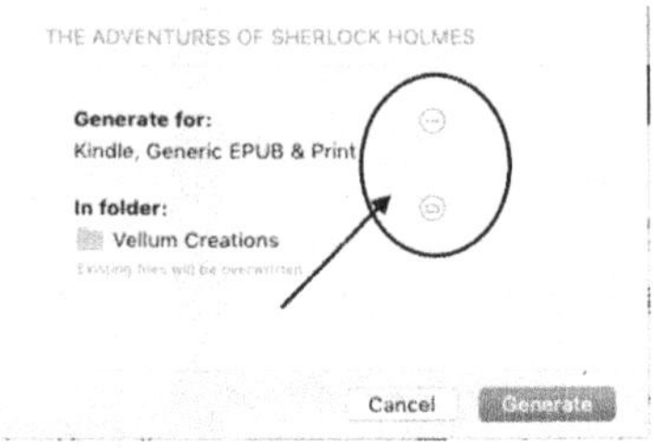

Fig. 62: Selectors for file types and saving location

Note: As mentioned in the "Files" chapter, you can create separate Vellum project files for ebooks and print books. Here is where you need to be sure you are generating the correct *types* of book files—ebooks or print. In the options shown in *Fig. 63*, select only "Print" if you are working in a print-specific file—do not select any ebook platforms. And vice versa—in your ebook file, select only the desired ebook platforms, but not print.

Fig. 63: Choose file types and stores

If you need to update your trim size, use the Print Settings menu (see the "Print Settings" chapter).

The lower circle from the popup window with the tiny folder icon (*Fig. 62*) lets you choose where your vellum files will be saved—in my case, "Vellum Creations" in my Dropbox folder (*Fig. 64*).

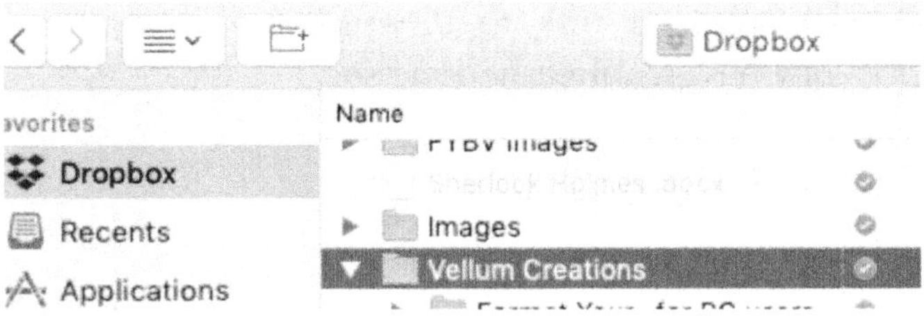

Fig. 64: Specify where to save your file

You can create a custom location with the New Folder button at the bottom left of this file location window or use an existing folder. Click Choose at the bottom right.

With these steps completed . . . it's time! Click the highlighted button at the bottom right that says Generate (*Fig. 65*).

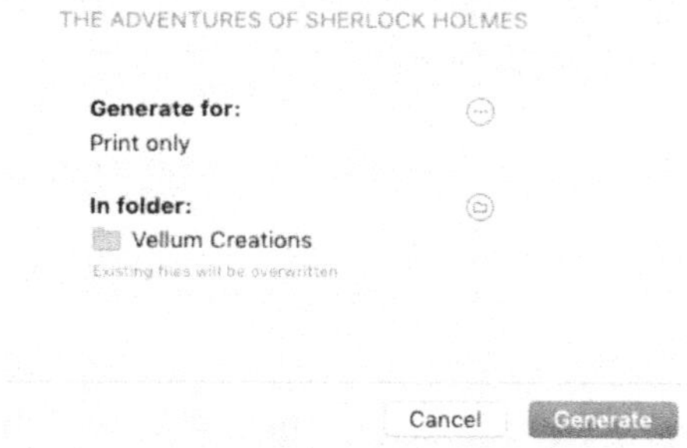

Fig. 65: Generate!

The process takes a few seconds, but when you see this window that says "Generation complete!" (*Fig. 66*), then you know you've been successful.

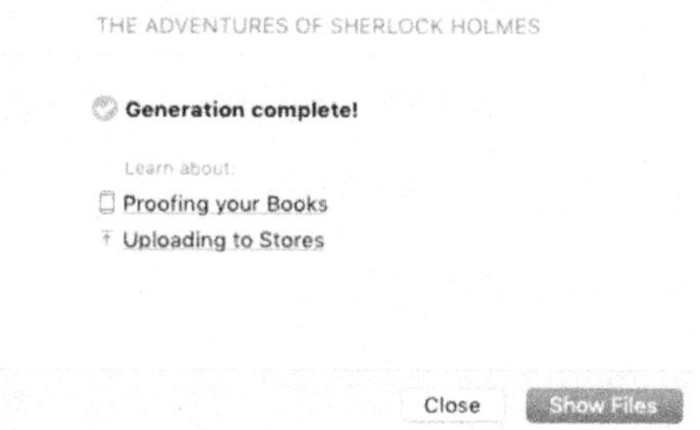

Fig. 66: Generation complete

Let's view those files, shall we?

Click Show Files (*Fig 66*).

Vellum has now created a subfolder *inside* my assigned folder location—Vellum Creations—and the folder is named the full book title, based on the My Book metadata (in the "Setup" chapter).

Now, inside the book-title file folder, each unique file is then saved into yet another subfolder by type (*Fig. 67*).

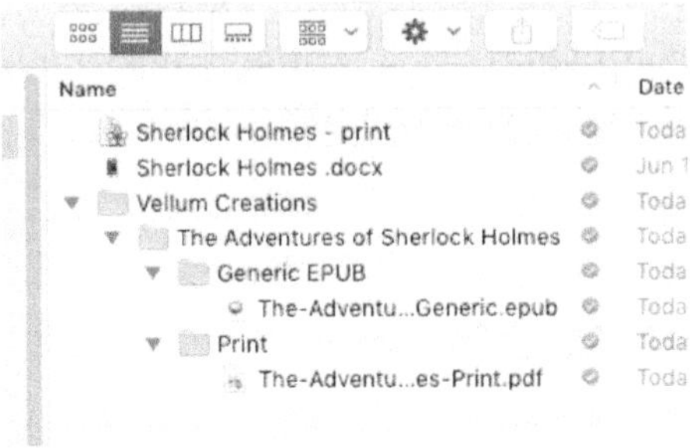

Fig. 67: Location of generated files

Whenever you generate books, the new file will be saved with the same basic name, *overwriting any identical files*—so if you want to keep multiple versions, you will need to rename what you just generated or drag/save it to a different file location. (For a quick rename without moving, click twice on the filename—separate clicks, not one double-click—to type directly onto it and rename. I find it helpful to include the date or another identifying factor so I know for sure which version I'm opening.)

CONGRATS!
You've successfully completed a start-to-finish book format!

VIEWING FILES

From here, you can view your files and see how they look. Your print file can be viewed on any PDF reader; Adobe Acrobat Reader and Foxit are two free options for either Mac or PC.

For EPUB viewing, there's no need to download any other software to your Mac. Using the Finder, locate the Generic EPUB file and click just once to select it. Go directly to the Menu Bar and click File > Open With > Books. There you go!

KINDLE AND INGRAMSPARK ERRORS

A note about Kindle files: Older versions of Vellum generate Kindle/MOBI files with text ornaments that may look too large when previewed on Kindle for Mac. (Ornaments display properly in purchased editions, just not in sideloaded files.) This issue was resolved in Vellum update 2.6.5. Users can upgrade to the newest version of Vellum to avoid this issue or view MOBI files using either the Kindle Previewer program or the "Launch Previewer" option after uploading to Kindle Direct Publishing.

IngramSpark error messages: Files uploaded to Ingram will show an error message about color profiles. You can ignore this and submit anyway, as this is a known issue.

SAVING A FINAL TEXT COPY

Some writers skip a word processing program and compose their book straight in Vellum. This can be helpful to see the project shaping up, but when they want to have their book edited, that is much better done in a program like Word so the writer can see the recommended changes.

Other users simply want a plain text copy of the final version.

While Vellum won't export a .docx file, it does support exporting to Rich Text Format (RTF). From the Menu Bar, click File > Export to RTF (*Fig. 68*). Choose where to store the file and click Save.

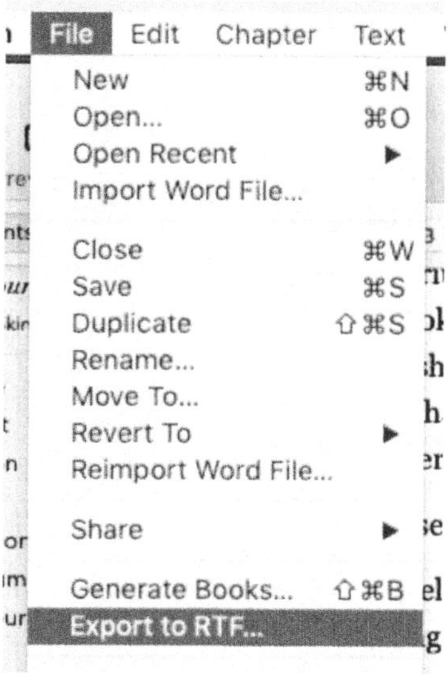

Fig. 68: Export to Rich Text Format (RTF)

An RTF document won't have images or fancy fine details, but it keeps all the basic text overrides like bold and italic. Sections like block quotations remain indented but do not have decorative items as in a Vellum book output. Headings remain separate from paragraphs, but the font is very simple.

To open or convert an RTF file, you can run your word processing program and then open the file. Make any desired changes and save as a .docx or other file format.

NOTES:

The process of dragging files into Vellum is more difficult to *describe* than it is to *do*—but making use of this unique feature can save you a great amount of time, effort, and decision-making. Not only can you quickly create coordinated books in a series or stick with settings that work well, you can use the same concept to efficiently update individual elements.

UPDATING A SINGLE ELEMENT

To update a single chapter or element in your book—like About the Author or an Also By page—create the desired information in one .docx file that can be inserted without needing to manually type in information or even copy / paste.

Remove the old element from the document; from the gear menu at the bottom of the Navigator, choose the Remove "[element]" option.

Open the Finder, choose the updated .docx file, and drag the file icon to the Navigator (*Fig. 69*).

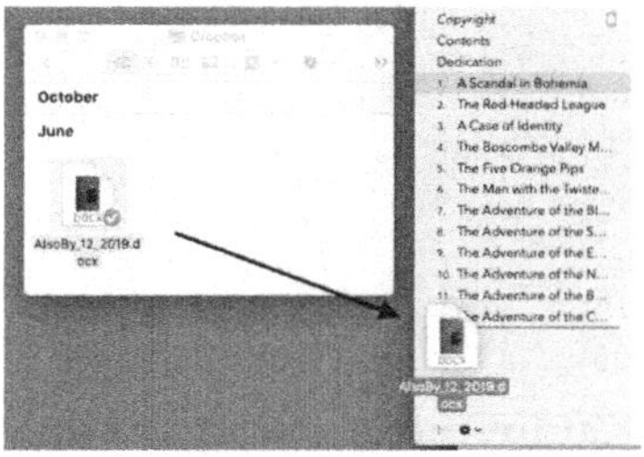

Fig. 69: Dragging a single .docx file into back matter

As you move the file, you'll see a line between elements indicating where the incoming file would be placed. Release the button, and the new section will be added! Rename and recategorize as necessary (see the "Elements" chapter), and tweak any headings or paragraphs. Done! Generate new files.

SETTING UP A VELLUM FILE AS A TEMPLATE

Duplicate existing document. To use an existing Vellum file as a template for another book, the first step is to copy the document. Open the file you previously set up. Then, from the Menu Bar click File > Duplicate. A new book file will automatically open.

Rename the duplicate. Click at the top of the new Vellum project and rename the copied file (*Fig. 70*). Hit Enter/Return to save.

Fig. 70: Rename the copied file as the new working filename

Update metadata. Click the old book title at the top of the Navigator pane (*Fig. 71*) and, under Title Info, update the title, subtitle, and any other information. Don't forget to update the copyright information or generate a replacement.

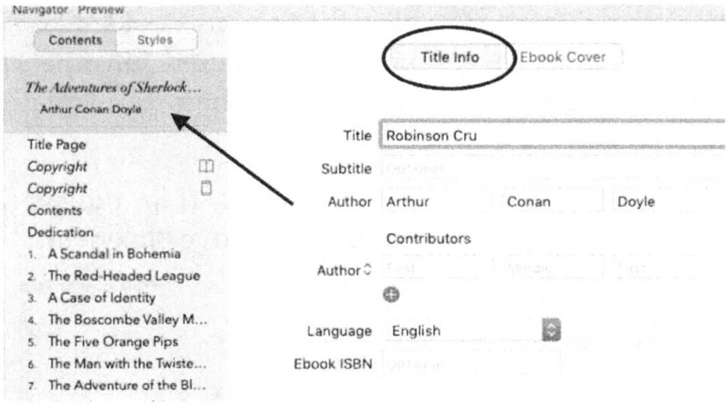

Fig. 71: Update metadata for new book

Remove the old cover image. If you added an ebook cover, you'll need to remove the old cover. Click Ebook Cover, then hover your cursor over the image while holding the Option key. This will convert the large "+" in the center (*Fig. 72*) to an "x," which will remove the image. (On PC, you can right-click to bring up the Remove/Replace option.) Then click the "+" and replace with a new image, if desired.

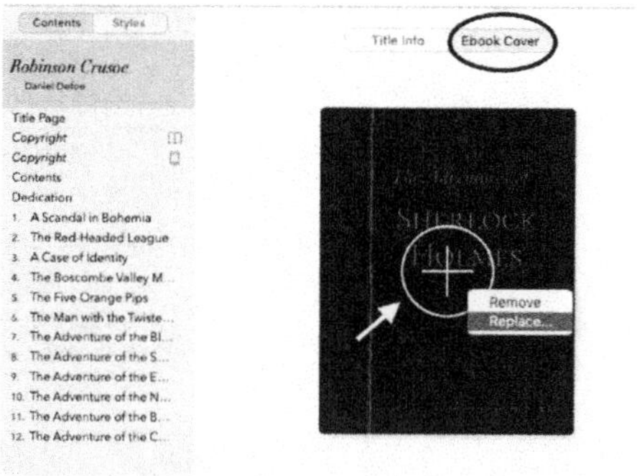

Fig. 72: Option + click to remove old ebook image

Remove old chapters. In the Navigator, click the first element or chapter that needs to be removed, then hold Shift and click the last element to select the entire group. From the bottom gear menu, choose Remove Selected Elements (*Fig. 73*).

Add the replacement book file. Now, locate your new .docx book file. Drag it directly onto the Navigator (*Fig. 74*). As you move the file, release the button where you want it placed. Make text adjustments as necessary and generate new files!

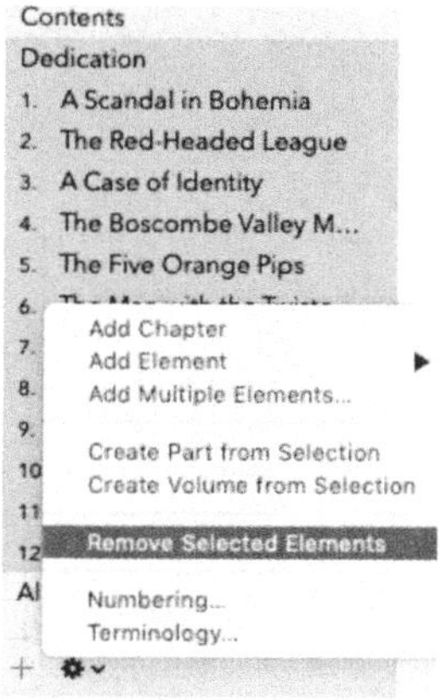

Fig. 73: Remove old material

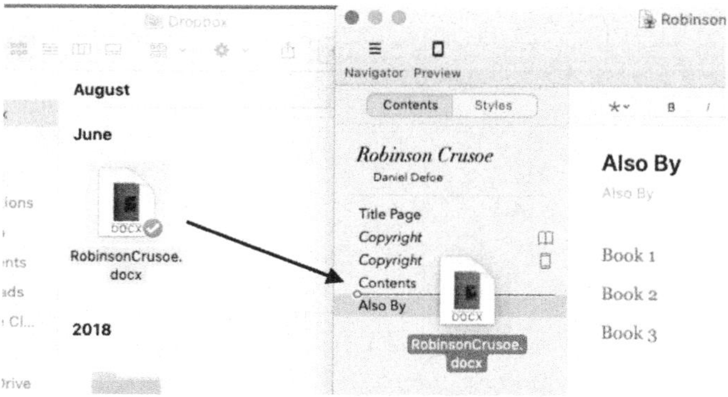

Fig. 74: Drag the new .docx file directly onto the template project

NOTES:

The ability to create multi-volume box sets is a brilliant feature by the 180g design team. Whether you're compiling your own collection or collaborating with other writers, this can be a fantastic item in your publishing arsenal.

BOX SETS USING WORD FILES

To create a box set using .docx files, create a new working project with a template file (see "Templates" chapter) or import a .docx file as a new project. Rename the working file and update the project metadata/title page information (see "Setup" chapter).

If you're using a template file, remove any unrelated elements (Navigator gear menu > Remove "[element]"). If you need to add the first book to the project, locate the .docx file in the Finder and drag it onto the Navigator to add it (see *Fig. 74* in previous chapter).

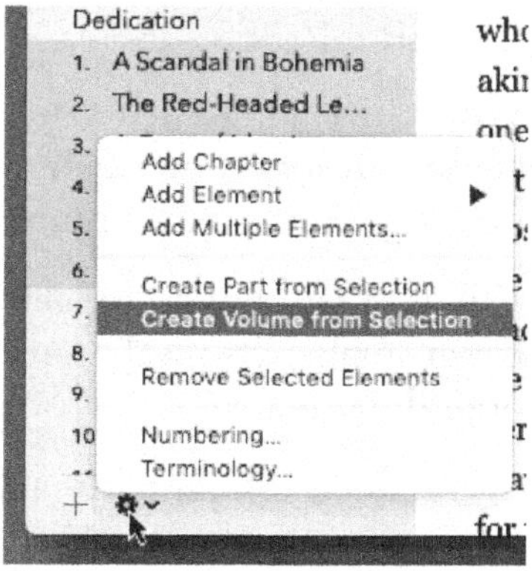

Fig. 75: Create Volume

Highlight all items for the new volume in the Navigator (shift + click will select the first and last), if they aren't selected automatically. Click the Navigator gear menu and choose Create Volume from Selection (*Fig. 75*).

Voilà! There's the first volume in your box set.

To add another file, choose it in the Finder. Drag the single .docx file onto the Navigator and place it where it needs to go in the book series. (Be careful to place it before or after the other volume—don't drag it into the same one. Edit > Undo is helpful if you do.) Now Shift + click to select all those chapters in the Navigator if needed, and again, choose Create Volume from Selection. Repeat as desired.

BOX SETS USING VELLUM FILES

Using existing Vellum book projects to compile box sets is even quicker than using Word files since each book is already styled and formatted.

If you start with an existing Vellum book file (a *copy* of the desired book, of course!), this will provide the style setup for the box set. Alternatively, open Vellum and choose File > New for a completely blank project file.

Use the same techniques as above to remove and insert books into the desired order. While Word files can be manually set as a new volume or dragged anywhere into an existing volume, Vellum files will always be their own volume, so you can drag in as many as you wish at one time. Drag and drop in the Navigator to reorder the volumes.

COVER IMAGES

While including each book's cover image is not a requirement (and some authors purposely don't include them to keep ebook file delivery size to a minimum), many readers enjoy seeing them.

If you create your box sets with Vellum files that include an ebook cover image, you will be prompted at import to choose the image or only the title text.

If you need to manually add a cover image, click on the first page of the desired volume. Choose the title gear menu and click Hide Heading in Book—but make sure your title still reads correctly, so it isn't only listed as a generic "Volume I" in the table of contents (TOC). Click down into the Text Editor and insert an image as the only thing on this page (see the "Text Features" chapter). Click on the image and set it at the largest size.

CONTENTS

As mentioned in the "Setup" chapter, online retailers typically require a "logical" table of contents programmed into ebooks, so Vellum automatically includes that regardless of whether you have the visible TOC included or not. But programming requirements aside, a fine-tuned TOC can help reader navigation, *especially* in box sets.

To add contents in a single volume, select Add Element > Table of Contents from the Navigator gear menu. The Contents element will generate near the top of the Navigator by default, so click and drag it just below the beginning of the desired volume (*Fig. 76*). Repeat for all volumes.

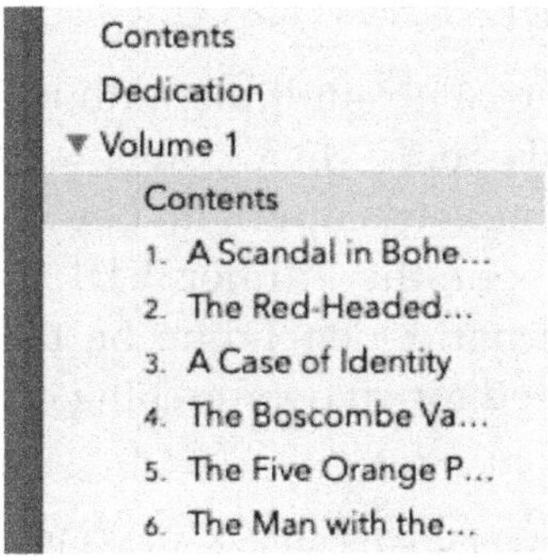

Fig. 76: Add contents to single volume only

To adjust the main book's overarching Table of Contents, make sure there is a Contents element before all the individual volumes. Select this to view the default contents list. If you want only the main book names to be listed and not the chapters, click Exclude chapters within volumes (*Fig. 77*).

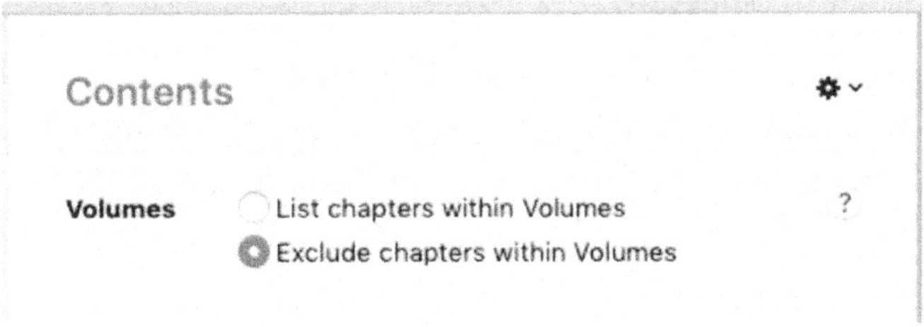

Fig. 77: Excluding chapters from main box set contents

SINGLE-AUTHOR BOX SETS

For a single-author box set, a full list of all book titles and their copyright year may be listed at the very beginning, with no need for further copyright information anywhere else in the set. Alternatively, this information can be more generic in the front, and the start of each volume can display the book-specific information of "Copyright © [year] by [author name]. All rights reserved." Either way, minimal non-story text in a box set streamlines the reading flow, especially for a series.

For back matter for one author, there is no need to include an Also By list or About the Author until the end of the entire set.

MULTI-AUTHOR BOX SETS

Sets with multiple authors can be more complex, but if you choose this route then you'll want to be sure every volume has its own complete copyright per that author, as well as their Also By and About the Author back matter segments in that volume. In this scenario, depending on the group arrangement, there may not be a need for an over-arching copyright statement before all the volumes.

PARTS WITHIN A VOLUME

If a book has Part I, Part II, etc., or if a work is made up of short stories that work best in parts, then it is simple to subdivide a volume. In the Navigator, select the all chapters in the volume that you wish to group, click the Navigator gear menu, and choose Create Part from Selection (*Fig. 78*).

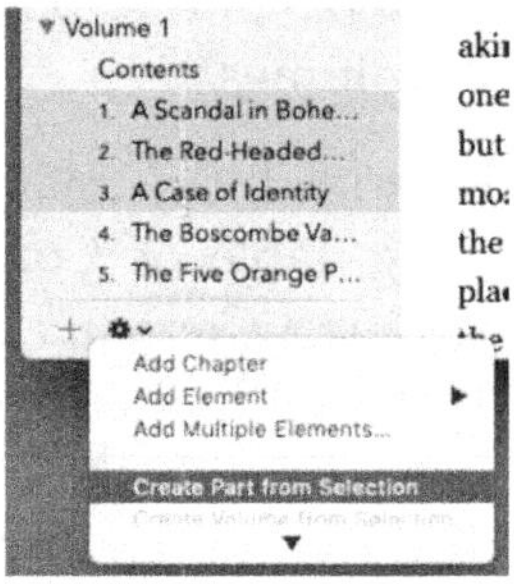

Fig. 78: Create parts within a volume

To adjust part or volume numbering, whether to restart numbering at 1 or make it continuous, click the Navigator gear menu and look at the Numbering options.

Author and podcaster Joanna Penn provides solid Vellum insights, including an informative article on the subject of box sets, which you can find here:

www.thecreativepenn.com/2016/07/06/create-box-set-bundle/

NOTES:

Learning Vellum *and* a new computer interface can be doubly troublesome. Here is a rundown of some basic Mac features and terms for my fellow PC natives.

MENU BAR

The Menu Bar is the thin strip across the top of your Mac screen (*Fig. 79*). This changes based on what program you are working in, as you can see by the name at the left end of the Menu Bar.

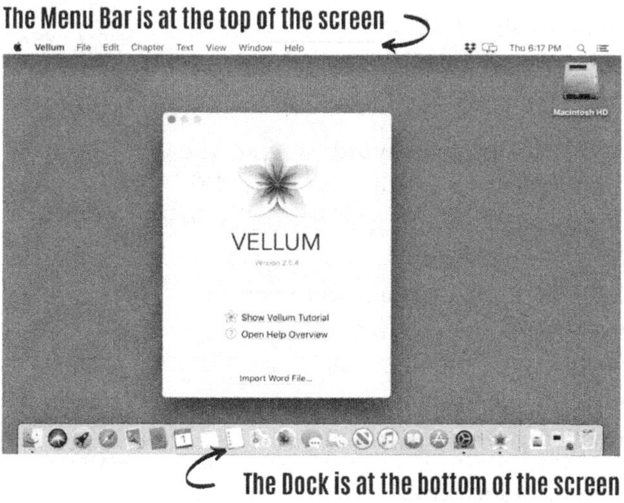

Fig. 79: Menu Bar and Dock

DOCK

The Dock is the toolbar across the bottom of the screen (*Fig. 79*) with default and recently used applications.

FINDER

Click on the Finder icon (*Fig. 80*) in the Dock to navigate to files saved on any connected drive location.

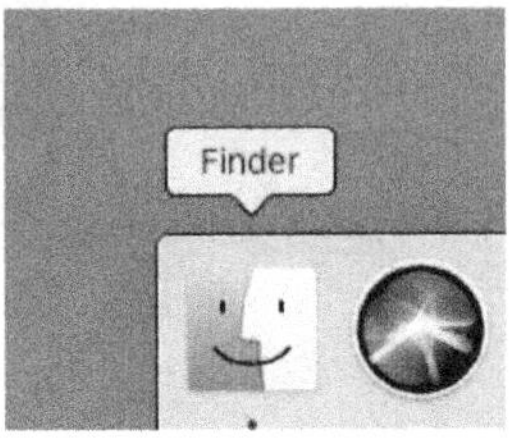

Fig. 80: Finder icon

LAUNCHPAD

The Launchpad icon (*Fig. 81*) opens a window which displays all the installed programs on your hard drive or remote server.

Fig. 81: Launchpad icon

APPLICATION WINDOWS

When you open a program, it will open in a rectangular area called a "window." The three colored circles—red, yellow, and green—at the top left corner of the window (*Fig. 82*) will resize or close the window.

Fig. 82: Top left of an application window: red/yellow/green circles

Red closes the application window, like the "x" at the top right of a window on a PC. ***Yellow*** minimizes it, sending it to the right side of the Dock, where you can click to reopen it. And ***green*** toggles between full screen and a smaller zoom view. (If your current window is at full size, click the green to toggle it to the smaller window, then minimize with yellow.)

Fig. 83: Dot below an application means program is running in the background

If there are small black dots below an icon in the Dock (*Fig. 83*), it means that program is running in the background, even if you closed the application window. Closing extra programs should help older machines and MacinCloud run a bit more efficiently.

If I wanted to completely shut down Vellum, for example, I would single-click on the Vellum Dock icon or click anywhere on an active Vellum window. Then, in the Menu Bar, click on the word *Vellum* and then scroll down to choose Quit Vellum (*Fig. 84*). Now the window is closed *and* the application is not running.

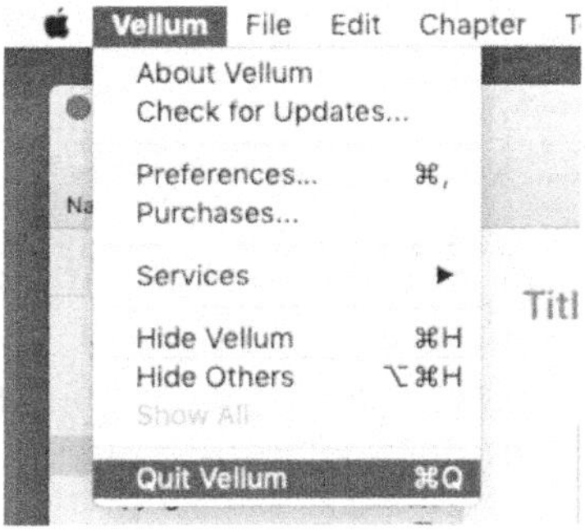

Fig. 84: Quit Vellum

KEYBOARD SHORTCUTS

When I'm accessing Vellum via MacinCloud, Mac keyboard shortcuts don't always translate well from a PC keyboard, so I use Edit in the Menu Bar. The Undo and Redo functions are at the top (and Undo is a good option for those "Oh no, I didn't mean to do *that*" moments).

To cut or copy and paste text on MacinCloud, I highlight the text, then either go to Edit > Copy/Cut or right-click and choose the option I want.

On a dedicated Mac, though, the Command key replaces the Control key used with common Windows shortcuts. The quick *undo* command for Mac is Cmd + Z, comparable to Ctrl + Z in Windows. *Redo* is Cmd + Shift + Z.

Similarly, the Windows keyboard shortcut commands for *copy* and *cut* are Ctrl + C and Ctrl + X, while on Mac they are Cmd + C and Cmd + X. *Paste* is Cmd + V.

VIEWING FILES AS ICONS, LISTS, ETC.

In my Sherlock Holmes sample project, some screenshots show files as folder icons in a grid and others as lists. To change this view in the Finder, open the Finder and find the Show Items As buttons (*Fig. 85*). The display options are icons, list, columns, or gallery.

Fig. 85: File viewing options: icons, list, columns, or gallery view

You can change the sorting or grouping of items a number of ways. In the Finder, locate Change Item Grouping (Fig. 86). Choose your desired sorting display.

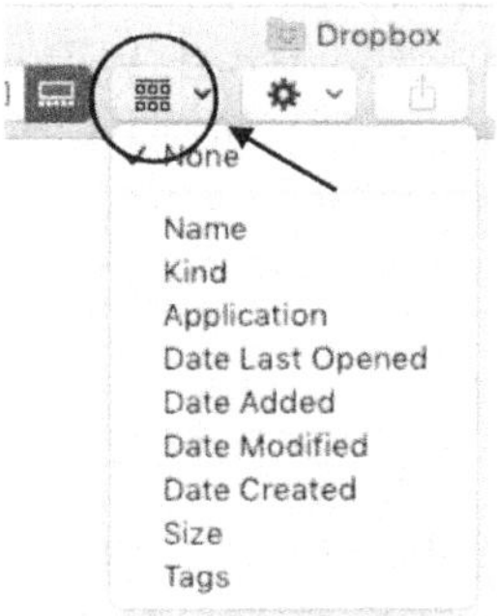

Fig. 86: Change file display grouping

SHUTTING DOWN OR LOGGING OUT

In the Menu Bar, choose the Apple logo at the far left end. Scroll down and choose Shut Down (*Fig. 87*) to power down your computer. For MacinCloud, as we'll explore in the next chapter, click the Apple logo and choose Log Out [username].

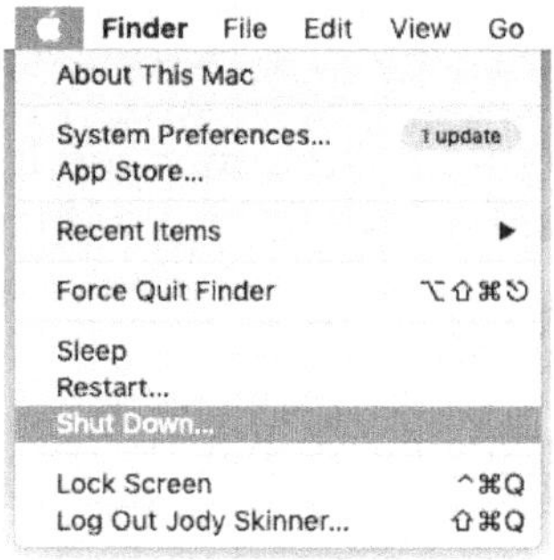

Fig. 87: Apple menu—shut down or log out

For more Mac information that isn't covered here, you may appreciate the helpful guide called "A Complete Beginner's Guide to MacOS."

www.makeuseof.com/tag/macos-beginners-guide/

NOTES:

14 MacinCloud

MacinCloud is a remote server resource which allows PC users to test out Vellum's features using their own computer.

Vellum is pre-installed on MacinCloud's managed servers, as is the Kindle Previewer software (necessary for generating book files for Kindle), so users can log in and immediately begin work on their project. Users simply need to purchase a Vellum license to be able to generate and export files, but all other program features are available to try. At present, MacinCloud provides the option of pay-as-you-go hours in thirty-hour chunks for $30 US, or there are weekly and monthly options starting at $12 and $20, respectively.

Just a little back story from me—I tested Vellum with MacinCloud's pay-as-you-go setup. I did forget to log out of the server once and "lost" about six purchased hours (logging out isn't terribly intuitive) —but you better believe I only forgot it once.

Navigating the Mac operating system wasn't too bad, but Vellum itself was maddening at first. I'd change my book style without realizing it, or I'd get confused as to where to find basic settings. When I poured almost $100 into MacinCloud hours in just a few weeks, I realized it wouldn't take long for a used Mac to pay for itself.

I was nervous about purchasing a used computer, but it was relatively simple to find a 2012 MacBook Pro for a few hundred dollars—and now over a year later, it's holding up well. (Don't tell my Dell—but I love the keyboard on my MacBook.) Many refurbished items carry a 90-day warranty, which, while not as good as a new computer's warranty, does allow for significantly heavy testing to make sure the item in question is going to hold up. Sites like Swappa, Gazelle, Woot!, and even Amazon have open-box and refurbished units.

But if you choose MacinCloud, whether for short- or long-term use, here is a basic walk-through.

SIGNING UP WITH MACINCLOUD

To begin with a MacinCloud setup, go to MacinCloud.com. At the time of this writing, there is a Compare Plans button on the home page, or from their menu options you can choose Pricing > Compare Plans. To direct link to this page is checkout.macincloud.com (*Fig. 88*).

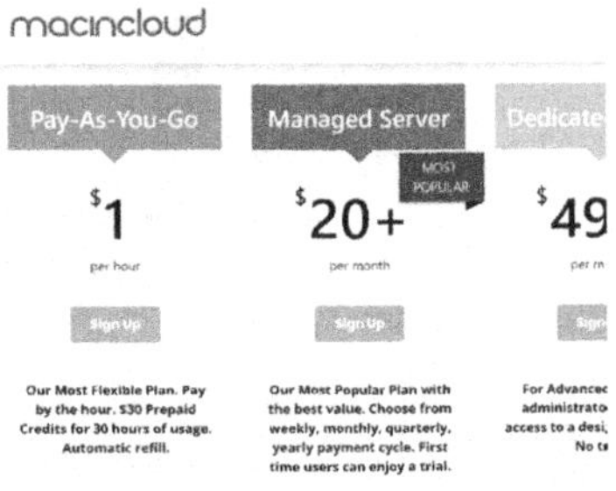

Fig. 88: Choosing a MacinCloud plan

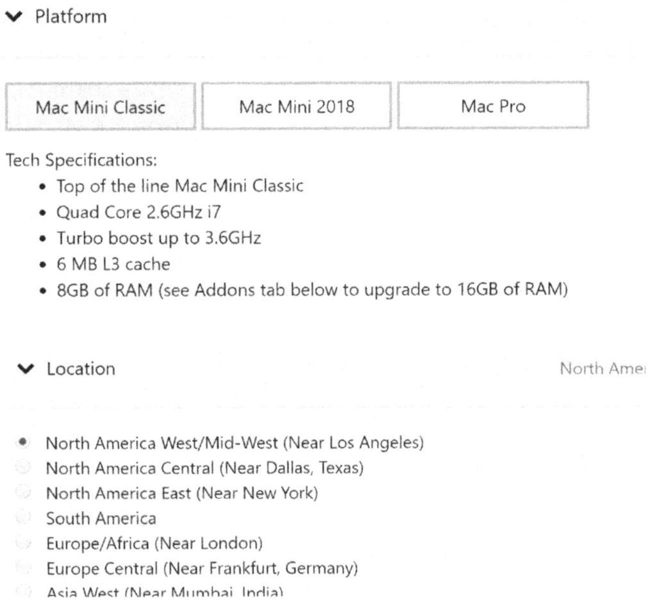

Fig. 89: Platform options and server locations

Click through and, at the top of the next page, choose any platform—Mac mini, etc. (*Fig. 89*). Any platform will run Vellum, but due to distance access, they do tend to run a bit slow. For server location, choose the nearest location to you.

For operating systems (*Fig. 90*), at the time of this printing, Vellum will run on Sierra, High Sierra, Mojave, and Catalina. Select billing frequency and desired daily access limit.

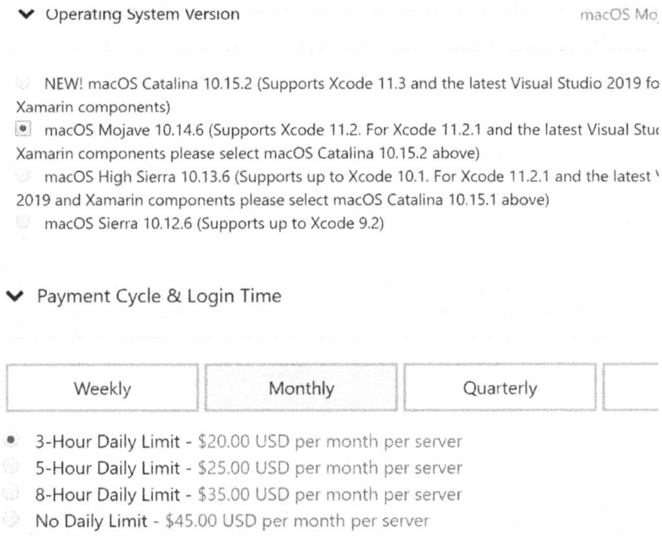

Fig. 90: Choose OS and subscription

These images are for the Managed Server options, but Pay-As-You-Go signup is very similar. Choose your preferred platform, the nearest server location, and the block of hours you wish to purchase, in $30-per-30-hour increments.

After you complete your purchase, you will receive your personal server information via email, along with an assigned login ID and a password to log in to your MacinCloud server.

ACCESS OPTION 1: MACINCLOUD IN YOUR BROWSER

Go to portal.macincloud.com and sign in (*Fig. 91*) using the email you used to purchase your server access. If you haven't yet created an account, use the account where your server purchase information was sent. This box is for general MacinCloud account access, not the server yet.

Signing in takes you directly to your dashboard (*Fig. 92*) where you can see your assigned server and subscription information. You are not using your paid time allowance until you connect to that server on the dashboard. To do that, click on your server name in the dashboard (in my case, "NY012"— *Fig. 92*) to go to the server page. Or select My Servers from the site menu bar and choose Managed, Dedicated, or Pay-As-You-Go, whichever you signed up for.

On your server page, find your server name and click Actions (*Fig. 93*). Choosing Connect Now will take you to the server login screen.

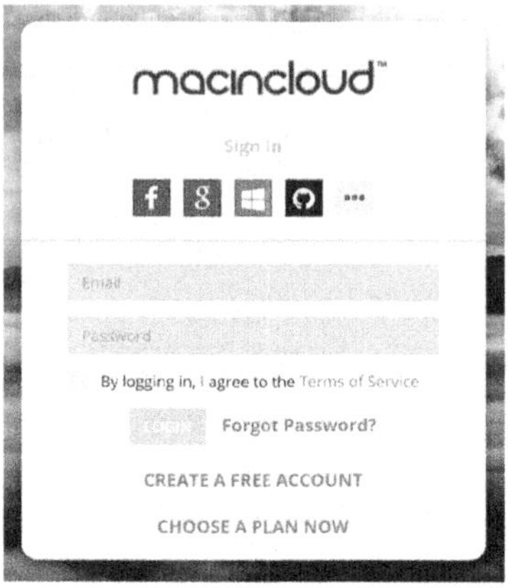

Fig. 91: MacinCloud site login—this is not your paid server access yet

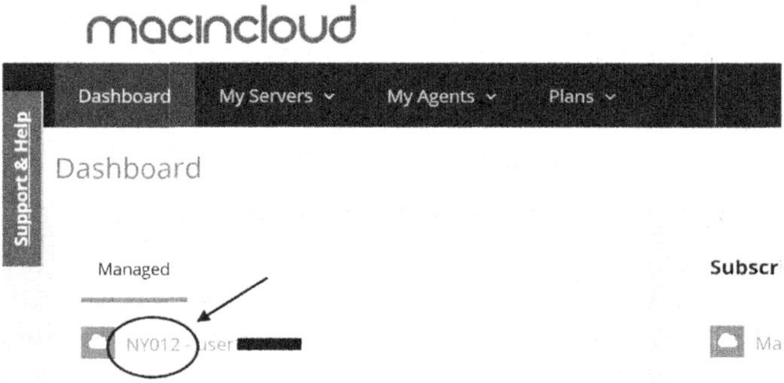

Fig. 92: MacinCloud dashboard

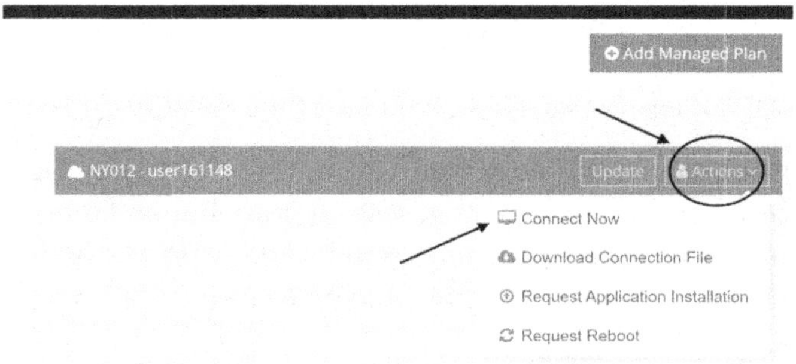

Fig. 93: Server—click Actions to connect

If you go to MacinCloud.com and your dashboard will not load, refresh the browser page and log in again.

ACCESS OPTION 2: DOWNLOADED CONNECTION FILE

Your registration from MacinCloud includes a link to download connection files for MacinCloud. This is a ZIP file with multiple sizes of pop-up windows that will open directly at your server login page. Download these and unzip the files if you wish to compare the view to your browser view or simply use this option instead.

You may have to test the browser and a few of the window sizes to see which size you prefer with your monitor display settings.

LOGGING IN: BROWSER OR POP-UP WINDOW

Whether you connect to MacinCloud through the online portal or you've opened one of the downloaded files, you now come to the screen where you can log in to your remote computer server (*Fig. 94*). Your login and password are in your welcome email from MacinCloud.

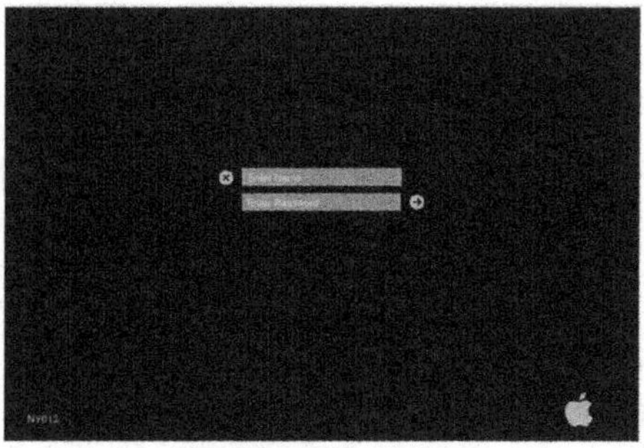

Fig. 94: Server login screen

Note: Because this is the first place you are encountering a separate computer, you cannot copy and paste text into the blanks—you will have to type in the information directly.

When you have successfully logged in, you will see your desktop screen. Note the Vellum flower logo in the Dock at the far right side

(*Fig. 95*)—it's pre-installed on managed and pay-as-you-go servers so you can get straight to work.

Fig. 95: MacinCloud desktop screen

Connect to Dropbox if you are using that to access your .docx files. At present, Dropbox is pre-installed on managed and pay-as-you-go servers.

Fig. 96: Launchpad and search blank at the top

If you don't see the Dropbox icon in your Dock, open the Launchpad and scroll through the downloaded apps or use the search blank at the top (*Fig. 96*). See the "Dropbox" chapter for download instructions, if needed.

Open Dropbox and connect your account. (See the "Dropbox" chapter for help.) From here, you can go to the "Files" chapter for instructions on file import.

MACINCLOUD AND ADOBE CREATIVE CLOUD

To access the Adobe Garamond Pro font using your active Adobe Creative Cloud subscription, you will need to request MacinCloud admin approval to install the Creative Cloud desktop app. You can download the installer file from www.adobe.com/creativecloud/desktop-app.html, then open a ticket at support.macincloud.com and enter your server number and username to request that they finalize the installation.

Once you receive confirmation that it is installed, you will see the Creative Cloud icon in the right side of the Menu Bar (*Fig. 97*). Click that small icon and log in with your Adobe information. If you have Adobe Garamond Pro activated in your Adobe Fonts, you will now see it as an option in the Body menu (see the "Styles" chapter).

Fig. 97: Adobe Creative Cloud icon —click to log in to Adobe

LOGGING OUT OF YOUR MACINCLOUD SERVER

Closing your browser tab or the application pop-up window ***does not log you out of your server!*** The clock is still ticking unless you complete the logout process.

To log out, click the Apple icon in the Menu Bar. Choose Log Out [user-name] (*Fig. 98*). Done!

Fig. 98: Log out of your MacinCloud server before closing the window or browser

If the application crashes, I log back in and do an official logout just to be safe. On one occasion, I couldn't log back in, so I immediately sent an email to support@macincloud.com with my server number and username to explain the situation and save my time allowance.

CANCELING A MACINCLOUD SUBSCRIPTION

If you know you don't want to continue your monthly subscription or have your Pay-As-You-Go hours automatically refilled, you can cancel your subscription and still access MacinCloud until your time limit is up or your hours are used.

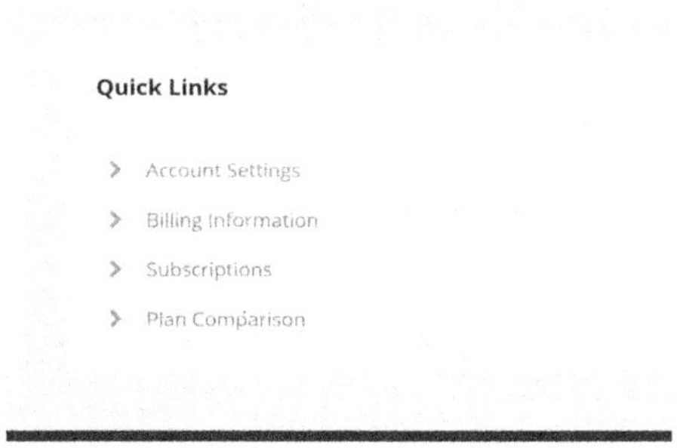

Fig. 99: Quick Links: chose Subscriptions to cancel

Log in to portal.MacinCloud.com. From the Dashboard, you can click scroll down to the Quick Links section and choose Subscriptions (*Fig. 99*) or click directly on the server name you wish to cancel in the Subscriptions box (*Fig. 100*).

Fig. 100: Click directly on server name you wish to cancel

To the right of your server name, click Update (*Fig. 101*).

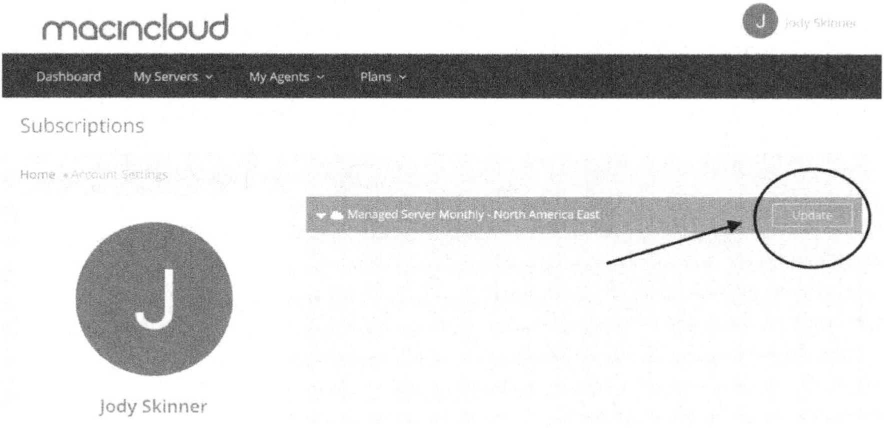

Fig. 101: Subscriptions area on MacinCloud portal

At the very bottom left corner of the pop-up window (*Fig. 102*), click Cancel Subscription and choose the appropriate options on the screens that follow.

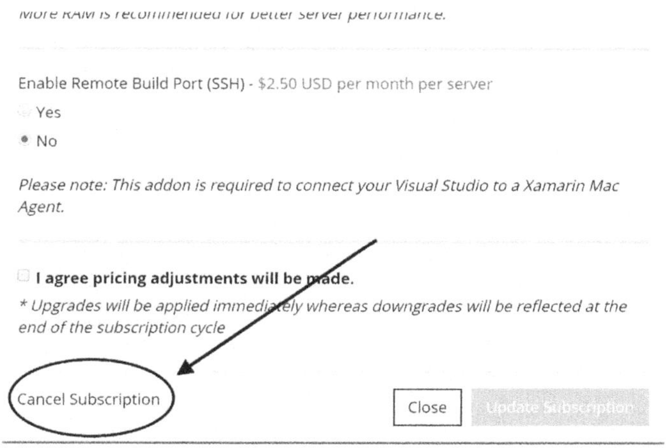

Fig. 102: Update subscriptions—Cancel Subscription at bottom left corner

NOTES:

Cloud file storage systems like Dropbox are a lifesaver for PC users, whether working on separate machines or using a single computer to access MacinCloud. While most cloud storage systems should work to connect your PC to your Mac or server, I'll walk you through Dropbox because it is already connected in MacinCloud. With its free pricing tier, which has 2GB of storage, you can connect up to three devices. (Even if you're only using one computer, your PC and your MacinCloud server would count as two devices).

These directions do assume you already have your .docx files saved to Dropbox for transfer. If you have not already signed up and completed this initial connection step, the process for new Dropbox users will walk you through their onboarding series.

CONNECTING A DROPBOX ACCOUNT

To connect to your Dropbox on a dedicated Mac computer, use a web browser to go to Dropbox.com and sign in.

Fig. 103: Downloads

A direct link to download the Dropbox application can be found at dropbox.com/downloading, if needed. When the download is complete, you'll see an icon drop to the right side of the dock into Downloads (*Fig. 103*).

Click the Downloads icon, select Dropbox Installer from the Downloads list (*Fig. 104*), and double-click the large Dropbox icon when prompted. (Click Open if a warning message comes up.)

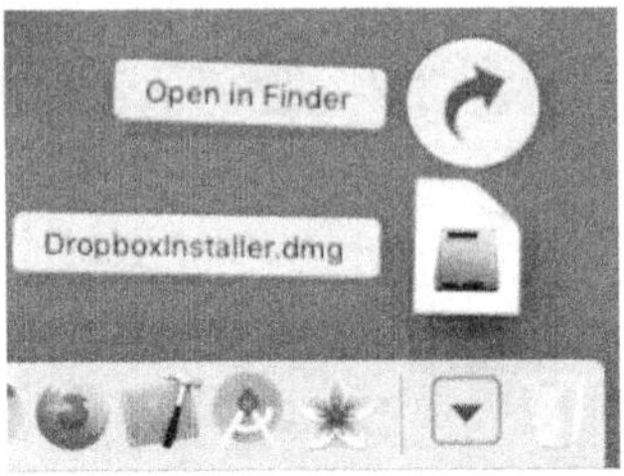

Fig. 104: Dropbox Installer—double-click to run

If you are connecting via MacinCloud in a browser, ensure you open Safari or another browser *on the server screen* to visit the download address—using your regular browser will download it to your PC, not to the server where you need it. See both address bars in *Fig. 105*. (It's a messy image. Sorry!)

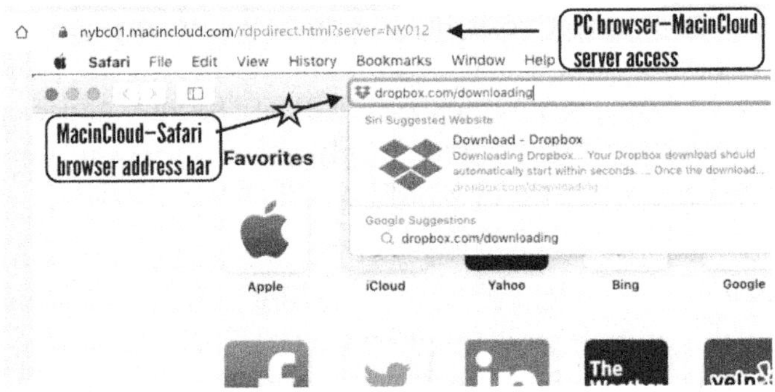

Fig. 105: PC browser address bar vs. MacinCloud browser address bar

You should now see a tiny Dropbox icon toward the right side of the Menu Bar (*Fig. 106*), and "Dropbox" will be an available save location in the Finder.

Fig. 106: Dropbox icon should be present in Menu Bar

If Dropbox did not automatically sign you in, click the small icon in the Menu Bar and it will take you through the sign in process, where you can click Open my Dropbox (*Fig. 107*) and connect to your account.

As long as you are signed in and have an internet connection, files will sync and be available from any access point.

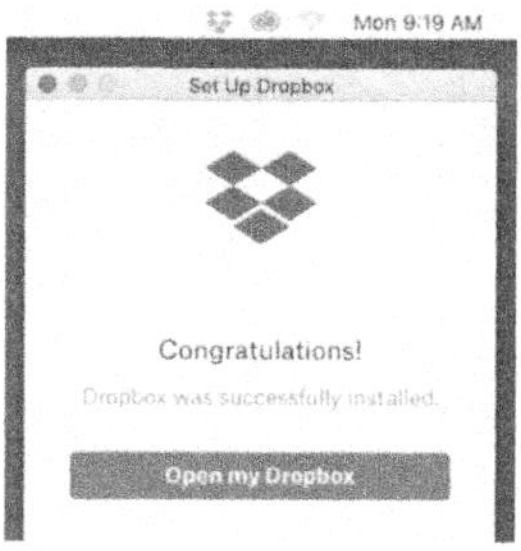

Fig. 107: Open my Dropbox

NOTES:

15 • DROPBOX

BEST WISHES ON YOUR BOOK FORMATTING JOURNEY!

So many words, and we still couldn't cover everything—but I sincerely hope all this information has been helpful to you, and that this resource boosts your confidence to successfully format your own book with Vellum. If you recognize that your project is beyond Vellum's scope and it's time to call a formatter, visit my free resource page at smidgenpress.pub/resources for a growing list of professionals and programs I personally recommend.

I look forward to seeing you and your book out in the big wide world of self-published authors. You'll find my info and social media links in "About the Author," and I'd love for you to tag me!

My best to you.
—Jody

PART TWO

VELLUM TEXT EXAMPLES

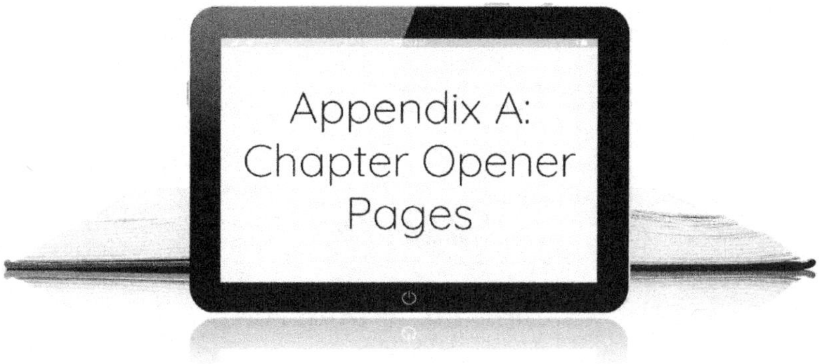

Appendix A contains all Vellum's basic chapter opener options. The pages are listed in alphabetical order by book style and designated #1, #2, etc., based on that particular Heading position option in the Style Carousel (indicated by the dots below the carousel options):

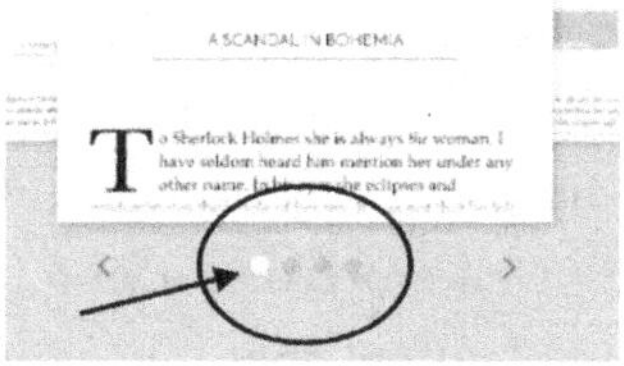

If you see a subtitle that lists a designation like "#1 & #4," that means the heading is identical for both options in the *print* view, but the *ebook* has multiple color options for that text position. The twenty-six options here in print are the default settings for the 6"x9" trim size, and you can view the heading options in the Style Carousel to see the ebook color variations.

Chapter One

A SCANDAL IN BOHEMIA

(Style: Artisanal | Heading #1)

To Sherlock Holmes she is always *the* woman. I have seldom heard him mention her under any other name. . . . Grit in a sensitive instrument, or a crack in one of his own high-power lenses, would not be more disturbing than a strong emotion in a nature such as his. And yet there was but one woman to him, and that woman was the late Irene Adler, of dubious and questionable memory.

And that was how a great scandal threatened to affect the kingdom of Bohemia, and how the best plans of Mr. Sherlock Holmes were beaten by a woman's wit. He used to make merry over the cleverness of women, but I have not heard him do it of late. And when he speaks of Irene Adler, or when he refers to her photograph, it is always under the honourable title of *the* woman.

1

A SCANDAL IN BOHEMIA

(Style: Artisanal | Heading #2)

To Sherlock Holmes she is always *the* woman. I have seldom heard him mention her under any other name. . . . Grit in a sensitive instrument, or a crack in one of his own high-power lenses, would not be more disturbing than a strong emotion in a nature such as his. And yet there was but one woman to him, and that woman was the late Irene Adler, of dubious and questionable memory.

And that was how a great scandal threatened to affect the kingdom of Bohemia, and how the best plans of Mr. Sherlock Holmes were beaten by a woman's wit. He used to make merry over the cleverness of women, but I have not heard him do it of late. And when he speaks of Irene Adler, or when he refers to her photograph, it is always under the honourable title of *the* woman.

A SCANDAL IN BOHEMIA

(Style: Artisanal | Heading #3)

A • CHAPTER OPENERS

To Sherlock Holmes she is always *the* woman. I have seldom heard him mention her under any other name. . . . Grit in a sensitive instrument, or a crack in one of his own high-power lenses, would not be more disturbing than a strong emotion in a nature such as his. And yet there was but one woman to him, and that woman was the late Irene Adler, of dubious and questionable memory.

And that was how a great scandal threatened to affect the kingdom of Bohemia, and how the best plans of Mr. Sherlock Holmes were beaten by a woman's wit. He used to make merry over the cleverness of women, but I have not heard him do it of late. And when he speaks of Irene Adler, or when he refers to her photograph, it is always under the honourable title of *the* woman.

CHAPTER 1

A SCANDAL IN BOHEMIA

(STYLE: CHROMA | HEADING #1 & #4)

TO SHERLOCK HOLMES she is always *the* woman. I have seldom heard him mention her under any other name. . . . Grit in a sensitive instrument, or a crack in one of his own high-power lenses, would not be more disturbing than a strong emotion in a nature such as his. And yet there was but one woman to him, and that woman was the late Irene Adler, of dubious and questionable memory.

And that was how a great scandal threatened to affect the kingdom of Bohemia, and how the best plans of Mr. Sherlock Holmes were beaten by a woman's wit. He used to make merry over the cleverness of women, but I have not heard him do it of late. And when he speaks of Irene Adler, or when he refers to her photograph, it is always under the honourable title of *the* woman.

CHAPTER ONE
A SCANDAL IN BOHEMIA

(STYLE: CHROMA | HEADING #2 & #5)

TO SHERLOCK HOLMES she is always *the* woman. I have seldom heard him mention her under any other name. . . . Grit in a sensitive instrument, or a crack in one of his own high-power lenses, would not be more disturbing than a strong emotion in a nature such as his. And yet there was but one woman to him, and that woman was the late Irene Adler, of dubious and questionable memory.

And that was how a great scandal threatened to affect the kingdom of Bohemia, and how the best plans of Mr. Sherlock Holmes were beaten by a woman's wit. He used to make merry over the cleverness of women, but I have not heard him do it of late. And when he speaks of Irene Adler, or when he refers to her photograph, it is always under the honourable title of *the* woman.

CHAPTER ONE

A SCANDAL IN BOHEMIA

(STYLE: CHROMA | HEADING #3 & #6)

TO SHERLOCK HOLMES she is always *the* woman. I have seldom heard him mention her under any other name. . . . Grit in a sensitive instrument, or a crack in one of his own high-power lenses, would not be more disturbing than a strong emotion in a nature such as his. And yet there was but one woman to him, and that woman was the late Irene Adler, of dubious and questionable memory.

And that was how a great scandal threatened to affect the kingdom of Bohemia, and how the best plans of Mr. Sherlock Holmes were beaten by a woman's wit. He used to make merry over the cleverness of women, but I have not heard him do it of late. And when he speaks of Irene Adler, or when he refers to her photograph, it is always under the honourable title of *the* woman.

A SCANDAL IN BOHEMIA

(STYLE: KINDRED | HEADING #1)

To Sherlock Holmes she is always *the* woman. I have seldom heard him mention her under any other name. . . . Grit in a sensitive instrument, or a crack in one of his own high-power lenses, would not be more disturbing than a strong emotion in a nature such as his. And yet there was but one woman to him, and that woman was the late Irene Adler, of dubious and questionable memory.

❄

And that was how a great scandal threatened to affect the kingdom of Bohemia, and how the best plans of Mr. Sherlock Holmes were beaten by a woman's wit. He used to make merry over the cleverness of women, but I have not heard him do it of late. And when he speaks of Irene Adler, or when he refers to her photograph, it is always under the honourable title of *the* woman.

1

A SCANDAL IN BOHEMIA

(STYLE: KINDRED | HEADING #2)

To Sherlock Holmes she is always *the* woman. I have seldom heard him mention her under any other name. . . . Grit in a sensitive instrument, or a crack in one of his own high-power lenses, would not be more disturbing than a strong emotion in a nature such as his. And yet there was but one woman to him, and that woman was the late Irene Adler, of dubious and questionable memory.

❄

AND THAT WAS how a great scandal threatened to affect the kingdom of Bohemia, and how the best plans of Mr. Sherlock Holmes were beaten by a woman's wit. He used to make merry over the cleverness of women, but I have not heard him do it of late. And when he speaks of Irene Adler, or when he refers to her photograph, it is always under the honourable title of *the* woman.

1

A SCANDAL IN BOHEMIA

(STYLE: KINDRED | HEADING #3)

To Sherlock Holmes she is always *the* woman. I have seldom heard him mention her under any other name. . . . Grit in a sensitive instrument, or a crack in one of his own high-power lenses, would not be more disturbing than a strong emotion in a nature such as his. And yet there was but one woman to him, and that woman was the late Irene Adler, of dubious and questionable memory.

❄

AND THAT WAS how a great scandal threatened to affect the kingdom of Bohemia, and how the best plans of Mr. Sherlock Holmes were beaten by a woman's wit. He used to make merry over the cleverness of women, but I have not heard him do it of late. And when he speaks of Irene Adler, or when he refers to her photograph, it is always under the honourable title of *the* woman.

A SCANDAL IN BOHEMIA

(STYLE: KINDRED | HEADING #4)

To Sherlock Holmes she is always *the* woman. I have seldom heard him mention her under any other name. . . . Grit in a sensitive instrument, or a crack in one of his own high-power lenses, would not be more disturbing than a strong emotion in a nature such as his. And yet there was but one woman to him, and that woman was the late Irene Adler, of dubious and questionable memory.

AND THAT WAS how a great scandal threatened to affect the kingdom of Bohemia, and how the best plans of Mr. Sherlock Holmes were beaten by a woman's wit. He used to make merry over the cleverness of women, but I have not heard him do it of late. And when he speaks of Irene Adler, or when he refers to her photograph, it is always under the honourable title of *the* woman.

1

A SCANDAL IN BOHEMIA

(STYLE: KINDRED | HEADING #5)

To Sherlock Holmes she is always *the* woman. I have seldom heard him mention her under any other name. . . . Grit in a sensitive instrument, or a crack in one of his own high-power lenses, would not be more disturbing than a strong emotion in a nature such as his. And yet there was but one woman to him, and that woman was the late Irene Adler, of dubious and questionable memory.

❄

AND THAT WAS how a great scandal threatened to affect the kingdom of Bohemia, and how the best plans of Mr. Sherlock Holmes were beaten by a woman's wit. He used to make merry over the cleverness of women, but I have not heard him do it of late. And when he speaks of Irene Adler, or when he refers to her photograph, it is always under the honourable title of *the* woman.

1

A SCANDAL IN BOHEMIA

(STYLE: MERIDIAN | HEADING #1)

To Sherlock Holmes she is always *the* woman. I have seldom heard him mention her under any other name. . . . Grit in a sensitive instrument, or a crack in one of his own high-power lenses, would not be more disturbing than a strong emotion in a nature such as his. And yet there was but one woman to him, and that woman was the late Irene Adler, of dubious and questionable memory.

And that was how a great scandal threatened to affect the kingdom of Bohemia, and how the best plans of Mr. Sherlock Holmes were beaten by a woman's wit. He used to make merry over the cleverness of women, but I have not heard him do it of late. And when he speaks of Irene Adler, or when he refers to her photograph, it is always under the honourable title of *the* woman.

Chapter 1

A SCANDAL IN BOHEMIA

(STYLE: MERIDIAN | HEADING #2)

To Sherlock Holmes she is always *the* woman. I have seldom heard him mention her under any other name. . . . Grit in a sensitive instrument, or a crack in one of his own high-power lenses, would not be more disturbing than a strong emotion in a nature such as his. And yet there was but one woman to him, and that woman was the late Irene Adler, of dubious and questionable memory.

And that was how a great scandal threatened to affect the kingdom of Bohemia, and how the best plans of Mr. Sherlock Holmes were beaten by a woman's wit. He used to make merry over the cleverness of women, but I have not heard him do it of late. And when he speaks of Irene Adler, or when he refers to her photograph, it is always under the honourable title of *the* woman.

1. A SCANDAL IN BOHEMIA

(STYLE: MERIDIAN | HEADING #3)

To Sherlock Holmes she is always *the* woman. I have seldom heard him mention her under any other name. . . . Grit in a sensitive instrument, or a crack in one of his own high-power lenses, would not be more disturbing than a strong emotion in a nature such as his. And yet there was but one woman to him, and that woman was the late Irene Adler, of dubious and questionable memory.

And that was how a great scandal threatened to affect the kingdom of Bohemia, and how the best plans of Mr. Sherlock Holmes were beaten by a woman's wit. He used to make merry over the cleverness of women, but I have not heard him do it of late. And when he speaks of Irene Adler, or when he refers to her photograph, it is always under the honourable title of *the* woman.

I

A SCANDAL IN BOHEMIA

(STYLE: OXFORD | HEADING #1)

To Sherlock Holmes she is always *the* woman. I have seldom heard him mention her under any other name. . . . Grit in a sensitive instrument, or a crack in one of his own high-power lenses, would not be more disturbing than a strong emotion in a nature such as his. And yet there was but one woman to him, and that woman was the late Irene Adler, of dubious and questionable memory.

And that was how a great scandal threatened to affect the kingdom of Bohemia, and how the best plans of Mr. Sherlock Holmes were beaten by a woman's wit. He used to make merry over the cleverness of women, but I have not heard him do it of late. And when he speaks of Irene Adler, or when he refers to her photograph, it is always under the honourable title of *the* woman.

A SCANDAL IN BOHEMIA

(STYLE: OXFORD | HEADING #2)

To Sherlock Holmes she is always *the* woman. I have seldom heard him mention her under any other name. . . . Grit in a sensitive instrument, or a crack in one of his own high-power lenses, would not be more disturbing than a strong emotion in a nature such as his. And yet there was but one woman to him, and that woman was the late Irene Adler, of dubious and questionable memory.

And that was how a great scandal threatened to affect the kingdom of Bohemia, and how the best plans of Mr. Sherlock Holmes were beaten by a woman's wit. He used to make merry over the cleverness of women, but I have not heard him do it of late. And when he speaks of Irene Adler, or when he refers to her photograph, it is always under the honourable title of *the* woman.

CHAPTER 1
A SCANDAL IN BOHEMIA
(STYLE: OXFORD | HEADING #3)

To Sherlock Holmes she is always *the* woman. I have seldom heard him mention her under any other name. . . . Grit in a sensitive instrument, or a crack in one of his own high-power lenses, would not be more disturbing than a strong emotion in a nature such as his. And yet there was but one woman to him, and that woman was the late Irene Adler, of dubious and questionable memory.

And that was how a great scandal threatened to affect the kingdom of Bohemia, and how the best plans of Mr. Sherlock Holmes were beaten by a woman's wit. He used to make merry over the cleverness of women, but I have not heard him do it of late. And when he speaks of Irene Adler, or when he refers to her photograph, it is always under the honourable title of *the* woman.

1

A SCANDAL IN BOHEMIA

(STYLE: PARCEL | HEADING #1)

To Sherlock Holmes she is always *the* woman. I have seldom heard him mention her under any other name. . . . Grit in a sensitive instrument, or a crack in one of his own high-power lenses, would not be more disturbing than a strong emotion in a nature such as his. And yet there was but one woman to him, and that woman was the late Irene Adler, of dubious and questionable memory.

~

And that was how a great scandal threatened to affect the kingdom of Bohemia, and how the best plans of Mr. Sherlock Holmes were beaten by a woman's wit. He used to make merry over the cleverness of women, but I have not heard him do it of late. And when he speaks of Irene Adler, or when he refers to her photograph, it is always under the honourable title of *the* woman.

1

A SCANDAL IN BOHEMIA

(STYLE: PARCEL | HEADING #2 & #3)

A • CHAPTER OPENERS

To Sherlock Holmes she is always *the* woman. I have seldom heard him mention her under any other name. . . . Grit in a sensitive instrument, or a crack in one of his own high-power lenses, would not be more disturbing than a strong emotion in a nature such as his. And yet there was but one woman to him, and that woman was the late Irene Adler, of dubious and questionable memory.

~

And that was how a great scandal threatened to affect the kingdom of Bohemia, and how the best plans of Mr. Sherlock Holmes were beaten by a woman's wit. He used to make merry over the cleverness of women, but I have not heard him do it of late. And when he speaks of Irene Adler, or when he refers to her photograph, it is always under the honourable title of *the* woman.

ONE

A SCANDAL IN BOHEMIA

(STYLE: PARCEL | HEADING #4)

To Sherlock Holmes she is always *the* woman. I have seldom heard him mention her under any other name. . . . Grit in a sensitive instrument, or a crack in one of his own high-power lenses, would not be more disturbing than a strong emotion in a nature such as his. And yet there was but one woman to him, and that woman was the late Irene Adler, of dubious and questionable memory.

And that was how a great scandal threatened to affect the kingdom of Bohemia, and how the best plans of Mr. Sherlock Holmes were beaten by a woman's wit. He used to make merry over the cleverness of women, but I have not heard him do it of late. And when he speaks of Irene Adler, or when he refers to her photograph, it is always under the honourable title of *the* woman.

1 / A SCANDAL IN BOHEMIA

(STYLE: SUDO | HEADING #1)

To Sherlock Holmes she is always *the* woman. I have seldom heard him mention her under any other name. . . . Grit in a sensitive instrument, or a crack in one of his own high-power lenses, would not be more disturbing than a strong emotion in a nature such as his. And yet there was but one woman to him, and that woman was the late Irene Adler, of dubious and questionable memory.

And that was how a great scandal threatened to affect the kingdom of Bohemia, and how the best plans of Mr. Sherlock Holmes were beaten by a woman's wit. He used to make merry over the cleverness of women, but I have not heard him do it of late. And when he speaks of Irene Adler, or when he refers to her photograph, it is always under the honourable title of *the* woman.

[1]

A SCANDAL IN BOHEMIA
(STYLE: SUDO | HEADING #2)

To Sherlock Holmes she is always *the* woman. I have seldom heard him mention her under any other name. . . . Grit in a sensitive instrument, or a crack in one of his own high-power lenses, would not be more disturbing than a strong emotion in a nature such as his. And yet there was but one woman to him, and that woman was the late Irene Adler, of dubious and questionable memory.

And that was how a great scandal threatened to affect the kingdom of Bohemia, and how the best plans of Mr. Sherlock Holmes were beaten by a woman's wit. He used to make merry over the cleverness of women, but I have not heard him do it of late. And when he speaks of Irene Adler, or when he refers to her photograph, it is always under the honourable title of *the* woman.

A SCANDAL IN BOHEMIA_
(STYLE: SUDO | HEADING #3)

To Sherlock Holmes she is always *the* woman. I have seldom heard him mention her under any other name. . . . Grit in a sensitive instrument, or a crack in one of his own high-power lenses, would not be more disturbing than a strong emotion in a nature such as his. And yet there was but one woman to him, and that woman was the late Irene Adler, of dubious and questionable memory.

And that was how a great scandal threatened to affect the kingdom of Bohemia, and how the best plans of Mr. Sherlock Holmes were beaten by a woman's wit. He used to make merry over the cleverness of women, but I have not heard him do it of late. And when he speaks of Irene Adler, or when he refers to her photograph, it is always under the honourable title of *the* woman.

ONE

A SCANDAL IN BOHEMIA

(STYLE: TRACE | HEADING #1)

TO SHERLOCK HOLMES she is always *the* woman. I have seldom heard him mention her under any other name. . . . Grit in a sensitive instrument, or a crack in one of his own high-power lenses, would not be more disturbing than a strong emotion in a nature such as his. And yet there was but one woman to him, and that woman was the late Irene Adler, of dubious and questionable memory.

And that was how a great scandal threatened to affect the kingdom of Bohemia, and how the best plans of Mr. Sherlock Holmes were beaten by a woman's wit. He used to make merry over the cleverness of women, but I have not heard him do it of late. And when he speaks of Irene Adler, or when he refers to her photograph, it is always under the honourable title of *the* woman.

1

A SCANDAL IN BOHEMIA

(STYLE: TRACE | HEADING #2)

TO SHERLOCK HOLMES she is always *the* woman. I have seldom heard him mention her under any other name. . . . Grit in a sensitive instrument, or a crack in one of his own high-power lenses, would not be more disturbing than a strong emotion in a nature such as his. And yet there was but one woman to him, and that woman was the late Irene Adler, of dubious and questionable memory.

And that was how a great scandal threatened to affect the kingdom of Bohemia, and how the best plans of Mr. Sherlock Holmes were beaten by a woman's wit. He used to make merry over the cleverness of women, but I have not heard him do it of late. And when he speaks of Irene Adler, or when he refers to her photograph, it is always under the honourable title of *the* woman.

1 A SCANDAL IN BOHEMIA

(STYLE: TRACE | HEADING #3)

TO SHERLOCK HOLMES she is always *the* woman. I have seldom heard him mention her under any other name. . . . Grit in a sensitive instrument, or a crack in one of his own high-power lenses, would not be more disturbing than a strong emotion in a nature such as his. And yet there was but one woman to him, and that woman was the late Irene Adler, of dubious and questionable memory.

And that was how a great scandal threatened to affect the kingdom of Bohemia, and how the best plans of Mr. Sherlock Holmes were beaten by a woman's wit. He used to make merry over the cleverness of women, but I have not heard him do it of late. And when he speaks of Irene Adler, or when he refers to her photograph, it is always under the honourable title of *the* woman.

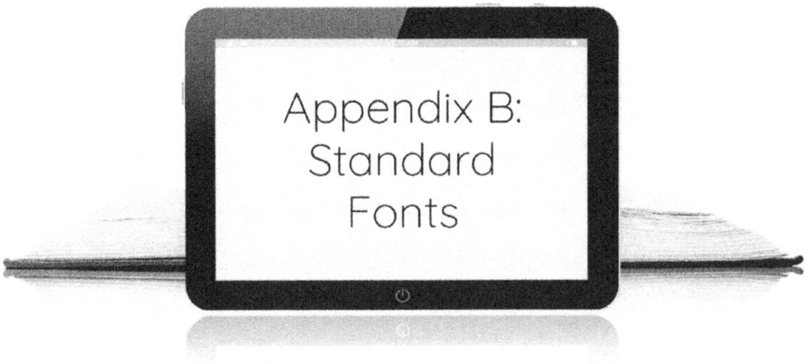

Well, here's the labor of love that started this whole book idea: the compilation of 140 text samples in this appendix and the next. As a PC user, my system didn't have any of the default Mac fonts used in Vellum, and I quickly grew tired of generating a test PDF, opening in whatever PDF reader I had, and trying to remember to "print without scaling" to see it at actual size. Somehow, it still wasn't quite working how I had hoped.

So here you go. This section and the following each have seventy font samples—standard font sizes in this section, large print in Appendix C. Each font has an introductory page with all English letter characters and numerals, based on the trim size of this book (6"x9").

Sizing: The samples in each section are numbered 1 through 7, according to the "Font Size" slider settings in the Print Settings dialog (File > Print Settings). I've included a representative slider image at the start of each section for visual learners, and on that I've also marked that all the samples are generated at the same "Line Spacing" slider setting (that's #2, in this section). Once you've determined the font size you want, you can manipulate the line spacing to take up more space or to conserve pages for a lower printing cost.

Every effort has been made to insert the samples at actual size, though with Vellum's Advanced Layout features, you might notice the line and paragraph spacing has been adjusted subtly in your project, depending on the text requirements. For more details, see help.vellum.pub/print/auto-layout.

ADOBE GARAMOND PRO*

ABCDEFGHIJKLMNOPQRSTUVWXYZ
abcdefghijklmnopqrstuvwxyz

ABCDEFGHIJKLMNOPQRSTUVWXYZ
abcdefghijklmnopqrstuvwxyz

ABCDEFGHIJKLMNOPQRSTUVWXYZ
abcdefghijklmnopqrstuvwxyz

ABCDEFGHIJKLMNOPQRSTUVWXYZ
abcdefghijklmnopqrstuvwxyz

0 1 2 3 4 5 6 7 8 9

*Only available if already installed/activated on your computer or server.

See "Styles" chapter for more details.

ADOBE GARAMOND PRO

Font sizes 1–7 | Line Spacing: 2

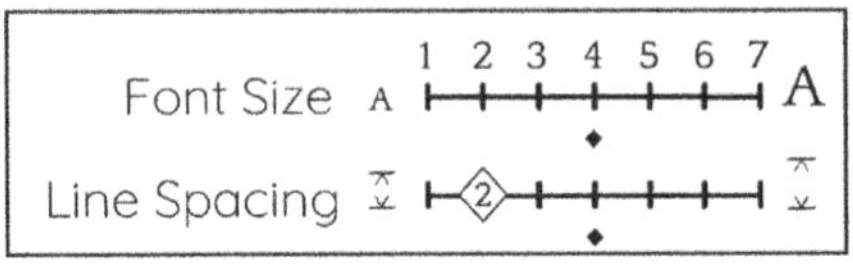

1. Does font choice really matter? Yes, actually—it might. Most readers can't express specifically why they do or don't like a book unless it *doesn't feel right,* so it's worth putting some thought into what you want to convey. Do you want an antique look for a classic retelling? See how your book reads in Baskerville. Do you simply prefer ease of readability with a professional appearance? Check out Iowan Old Style or Palatino and see what you think.

When you see how the words fall on these 6"x9" pages, you'll be better able to visualize your finished product. You'll confirm your own reading preferences or cater to those of your intended audience. Just remember—a primary goal is making the text so natural to read, so "invisible," that the reader forgets about individual words and instead enjoys the flow of the reading experience.

2. Does font choice really matter? Yes, actually—it might. Most readers can't express specifically why they do or don't like a book unless it *doesn't feel right,* so it's worth putting some thought into what you want to convey. Do you want an antique look for a classic retelling? See how your book reads in Baskerville. Do you simply prefer ease of readability with a professional appearance? Check out Iowan Old Style or Palatino and see what you think.

When you see how the words fall on these 6"x9" pages, you'll be better able to visualize your finished product. You'll confirm your own reading preferences or cater to those of your intended audience. Just remember—a primary goal is making the text so natural to read, so "invisible," that the reader forgets about individual words and instead enjoys the flow of the reading experience.

3. Does font choice really matter? Yes, actually—it might. Most readers can't express specifically why they do or don't like a book unless it *doesn't feel right,* so it's worth putting some thought into what you want to convey. Do you want an antique look for a classic retelling? See how your book reads in Baskerville. Do you simply prefer ease of readability with a professional appearance? Check out Iowan Old Style or Palatino and see what you think.

When you see how the words fall on these 6"x9" pages, you'll be better able to visualize your finished product. You'll confirm your own reading preferences or cater to those of your intended audience. Just remember—a primary goal is making the text so natural to read, so "invisible," that the reader forgets about individual words and instead enjoys the flow of the reading experience.

4. Does font choice really matter? Yes, actually—it might. Most readers can't express specifically why they do or don't like a book unless it *doesn't feel right*, so it's worth putting some thought into what you want to convey. Do you want an antique look for a classic retelling? See how your book reads in Baskerville. Do you simply prefer ease of readability with a professional appearance? Check out Iowan Old Style or Palatino and see what you think.

When you see how the words fall on these 6"x9" pages, you'll be better able to visualize your finished product. You'll confirm your own reading preferences or cater to those of your intended audience. Just remember—a primary goal is making the text so natural to read, so "invisible," that the reader forgets about individual words and instead enjoys the flow of the reading experience.

5. Does font choice really matter? Yes, actually—it might. Most readers can't express specifically why they do or don't like a book unless it *doesn't feel right*, so it's worth putting some thought into what you want to convey. Do you want an antique look for a classic retelling? See how your book reads in Baskerville. Do you simply prefer ease of readability with a professional appearance? Check out Iowan Old Style or Palatino and see what you think.

When you see how the words fall on these 6"x9" pages, you'll be better able to visualize your finished product. You'll confirm your own reading preferences or cater to those of your intended audience. Just remember—a primary goal is making the text so natural to read, so "invisible," that the reader forgets about individual words and instead enjoys the flow of the reading experience.

6. Does font choice really matter? Yes, actually—it might. Most readers can't express specifically why they do or don't like a book unless it *doesn't feel right*, so it's worth putting some thought into what you want to convey. Do you want an antique look for a classic retelling? See how your book reads in Baskerville. Do you simply prefer ease of readability with a professional appearance? Check out Iowan Old Style or Palatino and see what you think.

When you see how the words fall on these 6"x9" pages, you'll be better able to visualize your finished product. You'll confirm your own reading preferences or cater to those of your intended audience. Just remember—a primary goal is making the text so natural to read, so "invisible," that the reader forgets about individual words and instead enjoys the flow of the reading experience.

7. Does font choice really matter? Yes, actually—it might. Most readers can't express specifically why they do or don't like a book unless it *doesn't feel right*, so it's worth putting some thought into what you want to convey. Do you want an antique look for a classic retelling? See how your book reads in Baskerville. Do you simply prefer ease of readability with a professional appearance? Check out Iowan Old Style or Palatino and see what you think.

When you see how the words fall on these 6"x9" pages, you'll be better able to visualize your finished product. You'll confirm your own reading preferences or cater to those of your intended audience. Just remember—a primary goal is making the text so natural to read, so "invisible," that the reader forgets about individual words and instead enjoys the flow of the reading experience.

ATHELAS

ABCDEFGHIJKLMNOPQRSTUVWXYZ
abcdefghijklmnopqrstuvwxyz

ABCDEFGHIJKLMNOPQRSTUVWXYZ
abcdefghijklmnopqrstuvwxyz

ABCDEFGHIJKLMNOPQRSTUVWXYZ
abcdefghijklmnopqrstuvwxyz

ABCDEFGHIJKLMNOPQRSTUVWXYZ
abcdefghijklmnopqrstuvwxyz

0 1 2 3 4 5 6 7 8 9

ATHELAS

Font sizes 1–7 | Line Spacing: 2

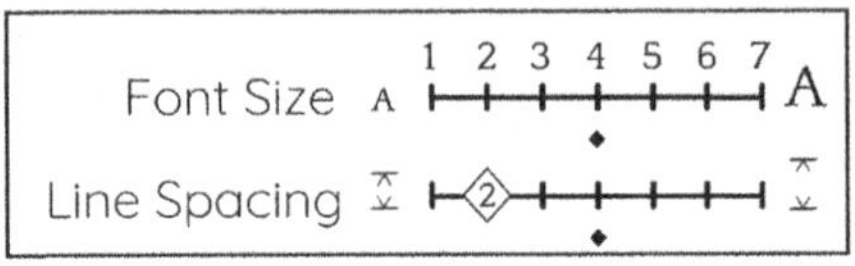

1. Does font choice really matter? Yes, actually—it might. Most readers can't express specifically why they do or don't like a book unless it *doesn't feel right*, so it's worth putting some thought into what you want to convey. Do you want an antique look for a classic retelling? See how your book reads in Baskerville. Do you simply prefer ease of readability with a professional appearance? Check out Iowan Old Style or Palatino and see what you think.

When you see how the words fall on these 6"x9" pages, you'll be better able to visualize your finished product. You'll confirm your own reading preferences or cater to those of your intended audience. Just remember—a primary goal is making the text so natural to read, so "invisible," that the reader forgets about individual words and instead enjoys the flow of the reading experience.

2. Does font choice really matter? Yes, actually—it might. Most readers can't express specifically why they do or don't like a book unless it *doesn't feel right*, so it's worth putting some thought into what you want to convey. Do you want an antique look for a classic retelling? See how your book reads in Baskerville. Do you simply prefer ease of readability with a professional appearance? Check out Iowan Old Style or Palatino and see what you think.

When you see how the words fall on these 6"x9" pages, you'll be better able to visualize your finished product. You'll confirm your own reading preferences or cater to those of your intended audience. Just remember—a primary goal is making the text so natural to read, so "invisible," that the reader forgets about individual words and instead enjoys the flow of the reading experience.

3. Does font choice really matter? Yes, actually—it might. Most readers can't express specifically why they do or don't like a book unless it *doesn't feel right*, so it's worth putting some thought into what you want to convey. Do you want an antique look for a classic retelling? See how your book reads in Baskerville. Do you simply prefer ease of readability with a professional appearance? Check out Iowan Old Style or Palatino and see what you think.

When you see how the words fall on these 6"x9" pages, you'll be better able to visualize your finished product. You'll confirm your own reading preferences or cater to those of your intended audience. Just remember—a primary goal is making the text so natural to read, so "invisible," that the reader forgets about individual words and instead enjoys the flow of the reading experience.

4. **Does font choice really matter?** Yes, actually—it might. Most readers can't express specifically why they do or don't like a book unless it *doesn't feel right*, so it's worth putting some thought into what you want to convey. Do you want an antique look for a classic retelling? See how your book reads in Baskerville. Do you simply prefer ease of readability with a professional appearance? Check out Iowan Old Style or Palatino and see what you think.

When you see how the words fall on these 6"x9" pages, you'll be better able to visualize your finished product. You'll confirm your own reading preferences or cater to those of your intended audience. Just remember—a primary goal is making the text so natural to read, so "invisible," that the reader forgets about individual words and instead enjoys the flow of the reading experience.

5. **Does font choice really matter?** Yes, actually—it might. Most readers can't express specifically why they do or don't like a book unless it *doesn't feel right*, so it's worth putting some thought into what you want to convey. Do you want an antique look for a classic retelling? See how your book reads in Baskerville. Do you simply prefer ease of readability with a professional appearance? Check out Iowan Old Style or Palatino and see what you think.

When you see how the words fall on these 6"x9" pages, you'll be better able to visualize your finished product. You'll confirm your own reading preferences or cater to those of your intended audience. Just remember—a primary goal is making the text so natural to read, so "invisible," that the reader forgets about individual words and instead enjoys the flow of the reading experience.

6. Does font choice really matter? Yes, actually—it might. Most readers can't express specifically why they do or don't like a book unless it *doesn't feel right*, so it's worth putting some thought into what you want to convey. Do you want an antique look for a classic retelling? See how your book reads in Baskerville. Do you simply prefer ease of readability with a professional appearance? Check out Iowan Old Style or Palatino and see what you think.

When you see how the words fall on these 6"x9" pages, you'll be better able to visualize your finished product. You'll confirm your own reading preferences or cater to those of your intended audience. Just remember—a primary goal is making the text so natural to read, so "invisible," that the reader forgets about individual words and instead enjoys the flow of the reading experience.

7. Does font choice really matter? Yes, actually—it might. Most readers can't express specifically why they do or don't like a book unless it *doesn't feel right*, so it's worth putting some thought into what you want to convey. Do you want an antique look for a classic retelling? See how your book reads in Baskerville. Do you simply prefer ease of readability with a professional appearance? Check out Iowan Old Style or Palatino and see what you think.

When you see how the words fall on these 6"x9" pages, you'll be better able to visualize your finished product. You'll confirm your own reading preferences or cater to those of your intended audience. Just remember—a primary goal is making the text so natural to read, so "invisible," that the reader forgets about individual words and instead enjoys the flow of the reading experience.

BASKERVILLE

ABCDEFGHIJKLMNOPQRSTUVWXYZ
abcdefghijklmnopqrstuvwxyz

ABCDEFGHIJKLMNOPQRSTUVWXYZ
abcdefghijklmnopqrstuvwxyz

ABCDEFGHIJKLMNOPQRSTUVWXYZ
abcdefghijklmnopqrstuvwxyz

ABCDEFGHIJKLMNOPQRSTUVWXYZ
abcdefghijklmnopqrstuvwxyz

0 1 2 3 4 5 6 7 8 9

BASKERVILLE
Font sizes 1–7 | Line Spacing: 2

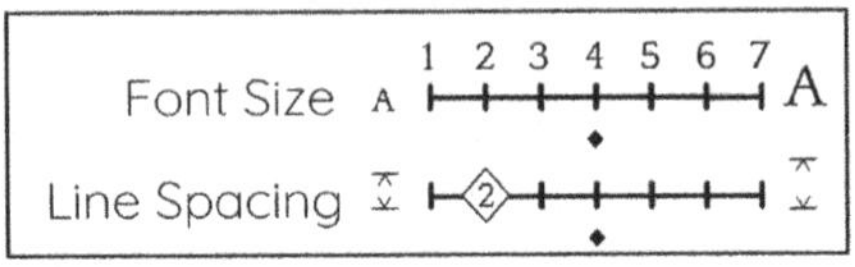

1. Does font choice really matter? Yes, actually—it might. Most readers can't express specifically why they do or don't like a book unless it *doesn't feel right*, so it's worth putting some thought into what you want to convey. Do you want an antique look for a classic retelling? See how your book reads in Baskerville. Do you simply prefer ease of readability with a professional appearance? Check out Iowan Old Style or Palatino and see what you think.

When you see how the words fall on these 6"x9" pages, you'll be better able to visualize your finished product. You'll confirm your own reading preferences or cater to those of your intended audience. Just remember—a primary goal is making the text so natural to read, so "invisible," that the reader forgets about individual words and instead enjoys the flow of the reading experience.

2. Does font choice really matter? Yes, actually—it might. Most readers can't express specifically why they do or don't like a book unless it *doesn't feel right*, so it's worth putting some thought into what you want to convey. Do you want an antique look for a classic retelling? See how your book reads in Baskerville. Do you simply prefer ease of readability with a professional appearance? Check out Iowan Old Style or Palatino and see what you think.

When you see how the words fall on these 6"x9" pages, you'll be better able to visualize your finished product. You'll confirm your own reading preferences or cater to those of your intended audience. Just remember—a primary goal is making the text so natural to read, so "invisible," that the reader forgets about individual words and instead enjoys the flow of the reading experience.

3. Does font choice really matter? Yes, actually—it might. Most readers can't express specifically why they do or don't like a book unless it *doesn't feel right*, so it's worth putting some thought into what you want to convey. Do you want an antique look for a classic retelling? See how your book reads in Baskerville. Do you simply prefer ease of readability with a professional appearance? Check out Iowan Old Style or Palatino and see what you think.

When you see how the words fall on these 6"x9" pages, you'll be better able to visualize your finished product. You'll confirm your own reading preferences or cater to those of your intended audience. Just remember—a primary goal is making the text so natural to read, so "invisible," that the reader forgets about individual words and instead enjoys the flow of the reading experience.

4. Does font choice really matter? Yes, actually—it might. Most readers can't express specifically why they do or don't like a book unless it *doesn't feel right,* so it's worth putting some thought into what you want to convey. Do you want an antique look for a classic retelling? See how your book reads in Baskerville. Do you simply prefer ease of readability with a professional appearance? Check out Iowan Old Style or Palatino and see what you think.

When you see how the words fall on these 6"x9" pages, you'll be better able to visualize your finished product. You'll confirm your own reading preferences or cater to those of your intended audience. Just remember—a primary goal is making the text so natural to read, so "invisible," that the reader forgets about individual words and instead enjoys the flow of the reading experience.

5. Does font choice really matter? Yes, actually—it might. Most readers can't express specifically why they do or don't like a book unless it *doesn't feel right,* so it's worth putting some thought into what you want to convey. Do you want an antique look for a classic retelling? See how your book reads in Baskerville. Do you simply prefer ease of readability with a professional appearance? Check out Iowan Old Style or Palatino and see what you think.

When you see how the words fall on these 6"x9" pages, you'll be better able to visualize your finished product. You'll confirm your own reading preferences or cater to those of your intended audience. Just remember—a primary goal is making the text so natural to read, so "invisible," that the reader forgets about individual words and instead enjoys the flow of the reading experience.

6. Does font choice really matter? Yes, actually—it might. Most readers can't express specifically why they do or don't like a book unless it *doesn't feel right*, so it's worth putting some thought into what you want to convey. Do you want an antique look for a classic retelling? See how your book reads in Baskerville. Do you simply prefer ease of readability with a professional appearance? Check out Iowan Old Style or Palatino and see what you think.

When you see how the words fall on these 6"x9" pages, you'll be better able to visualize your finished product. You'll confirm your own reading preferences or cater to those of your intended audience. Just remember—a primary goal is making the text so natural to read, so "invisible," that the reader forgets about individual words and instead enjoys the flow of the reading experience.

7. Does font choice really matter? Yes, actually—it might. Most readers can't express specifically why they do or don't like a book unless it *doesn't feel right*, so it's worth putting some thought into what you want to convey. Do you want an antique look for a classic retelling? See how your book reads in Baskerville. Do you simply prefer ease of readability with a professional appearance? Check out Iowan Old Style or Palatino and see what you think.

When you see how the words fall on these 6"x9" pages, you'll be better able to visualize your finished product. You'll confirm your own reading preferences or cater to those of your intended audience. Just remember—a primary goal is making the text so natural to read, so "invisible," that the reader forgets about individual words and instead enjoys the flow of the reading experience.

COCHIN

ABCDEFGHIJKLMNOPQRSTUVWXYZ
abcdefghijklmnopqrstuvwxyz

ABCDEFGHIJKLMNOPQRSTUVWXYZ
abcdefghijklmnopqrstuvwxyz

ABCDEFGHIJKLMNOPQRSTUVWXYZ
abcdefghijklmnopqrstuvwxyz

ABCDEFGHIJKLMNOPQRSTUVWXYZ
abcdefghijklmnopqrstuvwxyz

0 1 2 3 4 5 6 7 8 9

COCHIN

Font sizes 1–7 | Line Spacing: 2

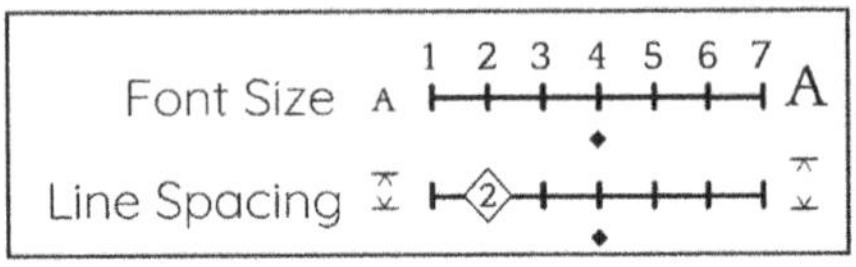

1. Does font choice really matter? Yes, actually—it might. Most readers can't express specifically why they do or don't like a book unless it *doesn't feel right,* so it's worth putting some thought into what you want to convey. Do you want an antique look for a classic retelling? See how your book reads in Baskerville. Do you simply prefer ease of readability with a professional appearance? Check out Iowan Old Style or Palatino and see what you think.

When you see how the words fall on these 6"x9" pages, you'll be better able to visualize your finished product. You'll confirm your own reading preferences or cater to those of your intended audience. Just remember—a primary goal is making the text so natural to read, so "invisible," that the reader forgets about individual words and instead enjoys the flow of the reading experience.

2. Does font choice really matter? Yes, actually—it might. Most readers can't express specifically why they do or don't like a book unless it *doesn't feel right,* so it's worth putting some thought into what you want to convey. Do you want an antique look for a classic retelling? See how your book reads in Baskerville. Do you simply prefer ease of readability with a professional appearance? Check out Iowan Old Style or Palatino and see what you think.

When you see how the words fall on these 6"x9" pages, you'll be better able to visualize your finished product. You'll confirm your own reading preferences or cater to those of your intended audience. Just remember—a primary goal is making the text so natural to read, so "invisible," that the reader forgets about individual words and instead enjoys the flow of the reading experience.

3. Does font choice really matter? Yes, actually—it might. Most readers can't express specifically why they do or don't like a book unless it *doesn't feel right,* so it's worth putting some thought into what you want to convey. Do you want an antique look for a classic retelling? See how your book reads in Baskerville. Do you simply prefer ease of readability with a professional appearance? Check out Iowan Old Style or Palatino and see what you think.

When you see how the words fall on these 6"x9" pages, you'll be better able to visualize your finished product. You'll confirm your own reading preferences or cater to those of your intended audience. Just remember—a primary goal is making the text so natural to read, so "invisible," that the reader forgets about individual words and instead enjoys the flow of the reading experience.

4. Does font choice really matter? Yes, actually—it might. Most readers can't express specifically why they do or don't like a book unless it *doesn't feel right,* so it's worth putting some thought into what you want to convey. Do you want an antique look for a classic retelling? See how your book reads in Baskerville. Do you simply prefer ease of readability with a professional appearance? Check out Iowan Old Style or Palatino and see what you think.

When you see how the words fall on these 6"x9" pages, you'll be better able to visualize your finished product. You'll confirm your own reading preferences or cater to those of your intended audience. Just remember—a primary goal is making the text so natural to read, so "invisible," that the reader forgets about individual words and instead enjoys the flow of the reading experience.

5. Does font choice really matter? Yes, actually—it might. Most readers can't express specifically why they do or don't like a book unless it *doesn't feel right,* so it's worth putting some thought into what you want to convey. Do you want an antique look for a classic retelling? See how your book reads in Baskerville. Do you simply prefer ease of readability with a professional appearance? Check out Iowan Old Style or Palatino and see what you think.

When you see how the words fall on these 6"x9" pages, you'll be better able to visualize your finished product. You'll confirm your own reading preferences or cater to those of your intended audience. Just remember—a primary goal is making the text so natural to read, so "invisible," that the reader forgets about individual words and instead enjoys the flow of the reading experience.

6. Does font choice really matter? Yes, actually—it might. Most readers can't express specifically why they do or don't like a book unless it *doesn't feel right,* so it's worth putting some thought into what you want to convey. Do you want an antique look for a classic retelling? See how your book reads in Baskerville. Do you simply prefer ease of readability with a professional appearance? Check out Iowan Old Style or Palatino and see what you think.

When you see how the words fall on these 6"x9" pages, you'll be better able to visualize your finished product. You'll confirm your own reading preferences or cater to those of your intended audience. Just remember—a primary goal is making the text so natural to read, so "invisible," that the reader forgets about individual words and instead enjoys the flow of the reading experience.

7. Does font choice really matter? Yes, actually—it might. Most readers can't express specifically why they do or don't like a book unless it *doesn't feel right,* so it's worth putting some thought into what you want to convey. Do you want an antique look for a classic retelling? See how your book reads in Baskerville. Do you simply prefer ease of readability with a professional appearance? Check out Iowan Old Style or Palatino and see what you think.

When you see how the words fall on these 6"x9" pages, you'll be better able to visualize your finished product. You'll confirm your own reading preferences or cater to those of your intended audience. Just remember—a primary goal is making the text so natural to read, so "invisible," that the reader forgets about individual words and instead enjoys the flow of the reading experience.

CRIMSON TEXT

ABCDEFGHIJKLMNOPQRSTUVWXYZ
abcdefghijklmnopqrstuvwxyz

ABCDEFGHIJKLMNOPQRSTUVWXYZ
abcdefghijklmnopqrstuvwxyz

ABCDEFGHIJKLMNOPQRSTUVWXYZ
abcdefghijklmnopqrstuvwxyz

ABCDEFGHIJKLMNOPQRSTUVWXYZ
abcdefghijklmnopqrstuvwxyz

0 1 2 3 4 5 6 7 8 9

CRIMSON TEXT
Font sizes 1–7 | Line Spacing: 2

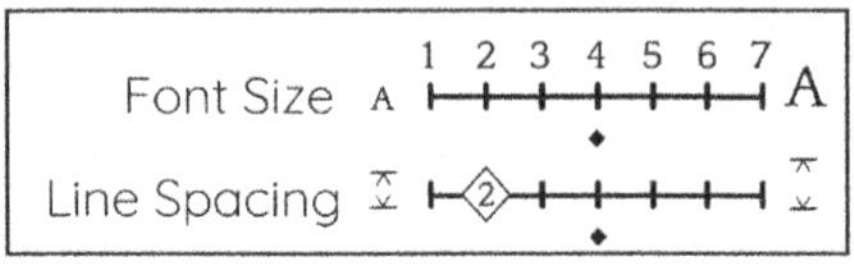

1. Does font choice really matter? Yes, actually—it might. Most readers can't express specifically why they do or don't like a book unless it *doesn't feel right*, so it's worth putting some thought into what you want to convey. Do you want an antique look for a classic retelling? See how your book reads in Baskerville. Do you simply prefer ease of readability with a professional appearance? Check out Iowan Old Style or Palatino and see what you think.

When you see how the words fall on these 6"x9" pages, you'll be better able to visualize your finished product. You'll confirm your own reading preferences or cater to those of your intended audience. Just remember—a primary goal is making the text so natural to read, so "invisible," that the reader forgets about individual words and instead enjoys the flow of the reading experience.

2. Does font choice really matter? Yes, actually—it might. Most readers can't express specifically why they do or don't like a book unless it *doesn't feel right*, so it's worth putting some thought into what you want to convey. Do you want an antique look for a classic retelling? See how your book reads in Baskerville. Do you simply prefer ease of readability with a professional appearance? Check out Iowan Old Style or Palatino and see what you think.

When you see how the words fall on these 6"x9" pages, you'll be better able to visualize your finished product. You'll confirm your own reading preferences or cater to those of your intended audience. Just remember—a primary goal is making the text so natural to read, so "invisible," that the reader forgets about individual words and instead enjoys the flow of the reading experience.

3. Does font choice really matter? Yes, actually—it might. Most readers can't express specifically why they do or don't like a book unless it *doesn't feel right*, so it's worth putting some thought into what you want to convey. Do you want an antique look for a classic retelling? See how your book reads in Baskerville. Do you simply prefer ease of readability with a professional appearance? Check out Iowan Old Style or Palatino and see what you think.

When you see how the words fall on these 6"x9" pages, you'll be better able to visualize your finished product. You'll confirm your own reading preferences or cater to those of your intended audience. Just remember—a primary goal is making the text so natural to read, so "invisible," that the reader forgets about individual words and instead enjoys the flow of the reading experience.

4. Does font choice really matter? Yes, actually—it might. Most readers can't express specifically why they do or don't like a book unless it *doesn't feel right,* so it's worth putting some thought into what you want to convey. Do you want an antique look for a classic retelling? See how your book reads in Baskerville. Do you simply prefer ease of readability with a professional appearance? Check out Iowan Old Style or Palatino and see what you think.

When you see how the words fall on these 6"x9" pages, you'll be better able to visualize your finished product. You'll confirm your own reading preferences or cater to those of your intended audience. Just remember—a primary goal is making the text so natural to read, so "invisible," that the reader forgets about individual words and instead enjoys the flow of the reading experience.

5. Does font choice really matter? Yes, actually—it might. Most readers can't express specifically why they do or don't like a book unless it *doesn't feel right,* so it's worth putting some thought into what you want to convey. Do you want an antique look for a classic retelling? See how your book reads in Baskerville. Do you simply prefer ease of readability with a professional appearance? Check out Iowan Old Style or Palatino and see what you think.

When you see how the words fall on these 6"x9" pages, you'll be better able to visualize your finished product. You'll confirm your own reading preferences or cater to those of your intended audience. Just remember—a primary goal is making the text so natural to read, so "invisible," that the reader forgets about individual words and instead enjoys the flow of the reading experience.

6. Does font choice really matter? Yes, actually—it might. Most readers can't express specifically why they do or don't like a book unless it *doesn't feel right,* so it's worth putting some thought into what you want to convey. Do you want an antique look for a classic retelling? See how your book reads in Baskerville. Do you simply prefer ease of readability with a professional appearance? Check out Iowan Old Style or Palatino and see what you think.

When you see how the words fall on these 6"x9" pages, you'll be better able to visualize your finished product. You'll confirm your own reading preferences or cater to those of your intended audience. Just remember—a primary goal is making the text so natural to read, so "invisible," that the reader forgets about individual words and instead enjoys the flow of the reading experience.

7. Does font choice really matter? Yes, actually—it might. Most readers can't express specifically why they do or don't like a book unless it *doesn't feel right,* so it's worth putting some thought into what you want to convey. Do you want an antique look for a classic retelling? See how your book reads in Baskerville. Do you simply prefer ease of readability with a professional appearance? Check out Iowan Old Style or Palatino and see what you think.

When you see how the words fall on these 6"x9" pages, you'll be better able to visualize your finished product. You'll confirm your own reading preferences or cater to those of your intended audience. Just remember—a primary goal is making the text so natural to read, so "invisible," that the reader forgets about individual words and instead enjoys the flow of the reading experience.

FANWOOD

ABCDEFGHIJKLMNOPQRSTUVWXYZ
abcdefghijklmnopqrstuvwxyz

ABCDEFGHIJKLMNOPQRSTUVWXYZ
abcdefghijklmnopqrstuvwxyz

ABCDEFGHIJKLMNOPQRSTUVWXYZ
abcdefghijklmnopqrstuvwxyz

ABCDEFGHIJKLMNOPQRSTUVWXYZ
abcdefghijklmnopqrstuvwxyz

0 1 2 3 4 5 6 7 8 9

FANWOOD

Font sizes 1–7 | Line Spacing: 2

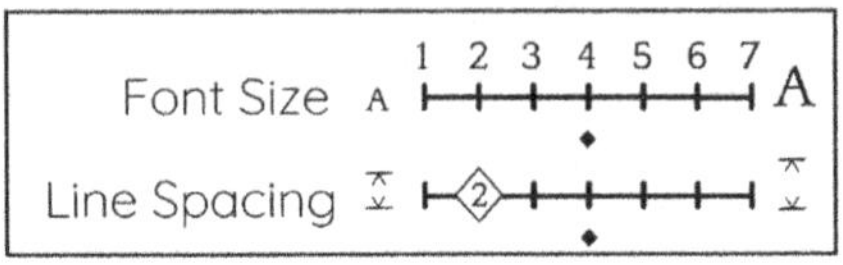

1. Does font choice really matter? Yes, actually—it might. Most readers can't express specifically why they do or don't like a book unless it *doesn't feel right*, so it's worth putting some thought into what you want to convey. Do you want an antique look for a classic retelling? See how your book reads in Baskerville. Do you simply prefer ease of readability with a professional appearance? Check out Iowan Old Style or Palatino and see what you think.

When you see how the words fall on these 6"x9" pages, you'll be better able to visualize your finished product. You'll confirm your own reading preferences or cater to those of your intended audience. Just remember—a primary goal is making the text so natural to read, so "invisible," that the reader forgets about individual words and instead enjoys the flow of the reading experience.

2. Does font choice really matter? Yes, actually—it might. Most readers can't express specifically why they do or don't like a book unless it *doesn't feel right*, so it's worth putting some thought into what you want to convey. Do you want an antique look for a classic retelling? See how your book reads in Baskerville. Do you simply prefer ease of readability with a professional appearance? Check out Iowan Old Style or Palatino and see what you think.

When you see how the words fall on these 6"x9" pages, you'll be better able to visualize your finished product. You'll confirm your own reading preferences or cater to those of your intended audience. Just remember—a primary goal is making the text so natural to read, so "invisible," that the reader forgets about individual words and instead enjoys the flow of the reading experience.

3. Does font choice really matter? Yes, actually—it might. Most readers can't express specifically why they do or don't like a book unless it *doesn't feel right*, so it's worth putting some thought into what you want to convey. Do you want an antique look for a classic retelling? See how your book reads in Baskerville. Do you simply prefer ease of readability with a professional appearance? Check out Iowan Old Style or Palatino and see what you think.

When you see how the words fall on these 6"x9" pages, you'll be better able to visualize your finished product. You'll confirm your own reading preferences or cater to those of your intended audience. Just remember—a primary goal is making the text so natural to read, so "invisible," that the reader forgets about individual words and instead enjoys the flow of the reading experience.

4. Does font choice really matter? Yes, actually—it might. Most readers can't express specifically why they do or don't like a book unless it *doesn't feel right*, so it's worth putting some thought into what you want to convey. Do you want an antique look for a classic retelling? See how your book reads in Baskerville. Do you simply prefer ease of readability with a professional appearance? Check out Iowan Old Style or Palatino and see what you think.

When you see how the words fall on these 6"x9" pages, you'll be better able to visualize your finished product. You'll confirm your own reading preferences or cater to those of your intended audience. Just remember—a primary goal is making the text so natural to read, so "invisible," that the reader forgets about individual words and instead enjoys the flow of the reading experience.

5. Does font choice really matter? Yes, actually—it might. Most readers can't express specifically why they do or don't like a book unless it *doesn't feel right*, so it's worth putting some thought into what you want to convey. Do you want an antique look for a classic retelling? See how your book reads in Baskerville. Do you simply prefer ease of readability with a professional appearance? Check out Iowan Old Style or Palatino and see what you think.

When you see how the words fall on these 6"x9" pages, you'll be better able to visualize your finished product. You'll confirm your own reading preferences or cater to those of your intended audience. Just remember—a primary goal is making the text so natural to read, so "invisible," that the reader forgets about individual words and instead enjoys the flow of the reading experience.

6. Does font choice really matter? Yes, actually—it might. Most readers can't express specifically why they do or don't like a book unless it *doesn't feel right*, so it's worth putting some thought into what you want to convey. Do you want an antique look for a classic retelling? See how your book reads in Baskerville. Do you simply prefer ease of readability with a professional appearance? Check out Iowan Old Style or Palatino and see what you think.

When you see how the words fall on these 6"x9" pages, you'll be better able to visualize your finished product. You'll confirm your own reading preferences or cater to those of your intended audience. Just remember—a primary goal is making the text so natural to read, so "invisible," that the reader forgets about individual words and instead enjoys the flow of the reading experience.

7. Does font choice really matter? Yes, actually—it might. Most readers can't express specifically why they do or don't like a book unless it *doesn't feel right*, so it's worth putting some thought into what you want to convey. Do you want an antique look for a classic retelling? See how your book reads in Baskerville. Do you simply prefer ease of readability with a professional appearance? Check out Iowan Old Style or Palatino and see what you think.

When you see how the words fall on these 6"x9" pages, you'll be better able to visualize your finished product. You'll confirm your own reading preferences or cater to those of your intended audience. Just remember—a primary goal is making the text so natural to read, so "invisible," that the reader forgets about individual words and instead enjoys the flow of the reading experience.

HOEFLER TEXT

ABCDEFGHIJKLMNOPQRSTUVWXYZ
abcdefghijklmnopqrstuvwxyz

ABCDEFGHIJKLMNOPQRSTUVWXYZ
abcdefghijklmnopqrstuvwxyz

ABCDEFGHIJKLMNOPQRSTUVWXYZ
abcdefghijklmnopqrstuvwxyz

ABCDEFGHIJKLMNOPQRSTUVWXYZ
abcdefghijklmnopqrstuvwxyz

0 1 2 3 4 5 6 7 8 9

HOEFLER TEXT

Font sizes 1–7 | Line Spacing: 2

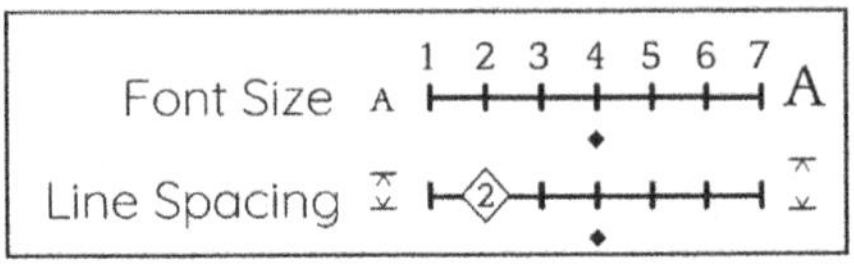

1. **Does font choice really matter?** Yes, actually—it might. Most readers can't express specifically why they do or don't like a book unless it *doesn't feel right*, so it's worth putting some thought into what you want to convey. Do you want an antique look for a classic retelling? See how your book reads in Baskerville. Do you simply prefer ease of readability with a professional appearance? Check out Iowan Old Style or Palatino and see what you think.

When you see how the words fall on these 6"x9" pages, you'll be better able to visualize your finished product. You'll confirm your own reading preferences or cater to those of your intended audience. Just remember—a primary goal is making the text so natural to read, so "invisible," that the reader forgets about individual words and instead enjoys the flow of the reading experience.

2. **Does font choice really matter?** Yes, actually—it might. Most readers can't express specifically why they do or don't like a book unless it *doesn't feel right*, so it's worth putting some thought into what you want to convey. Do you want an antique look for a classic retelling? See how your book reads in Baskerville. Do you simply prefer ease of readability with a professional appearance? Check out Iowan Old Style or Palatino and see what you think.

When you see how the words fall on these 6"x9" pages, you'll be better able to visualize your finished product. You'll confirm your own reading preferences or cater to those of your intended audience. Just remember—a primary goal is making the text so natural to read, so "invisible," that the reader forgets about individual words and instead enjoys the flow of the reading experience.

3. **Does font choice really matter?** Yes, actually—it might. Most readers can't express specifically why they do or don't like a book unless it *doesn't feel right*, so it's worth putting some thought into what you want to convey. Do you want an antique look for a classic retelling? See how your book reads in Baskerville. Do you simply prefer ease of readability with a professional appearance? Check out Iowan Old Style or Palatino and see what you think.

When you see how the words fall on these 6"x9" pages, you'll be better able to visualize your finished product. You'll confirm your own reading preferences or cater to those of your intended audience. Just remember—a primary goal is making the text so natural to read, so "invisible," that the reader forgets about individual words and instead enjoys the flow of the reading experience.

4. **Does font choice really matter?** Yes, actually—it might. Most readers can't express specifically why they do or don't like a book unless it *doesn't feel right*, so it's worth putting some thought into what you want to convey. Do you want an antique look for a classic retelling? See how your book reads in Baskerville. Do you simply prefer ease of readability with a professional appearance? Check out Iowan Old Style or Palatino and see what you think.

When you see how the words fall on these 6"x9" pages, you'll be better able to visualize your finished product. You'll confirm your own reading preferences or cater to those of your intended audience. Just remember—a primary goal is making the text so natural to read, so "invisible," that the reader forgets about individual words and instead enjoys the flow of the reading experience.

5. **Does font choice really matter?** Yes, actually—it might. Most readers can't express specifically why they do or don't like a book unless it *doesn't feel right*, so it's worth putting some thought into what you want to convey. Do you want an antique look for a classic retelling? See how your book reads in Baskerville. Do you simply prefer ease of readability with a professional appearance? Check out Iowan Old Style or Palatino and see what you think.

When you see how the words fall on these 6"x9" pages, you'll be better able to visualize your finished product. You'll confirm your own reading preferences or cater to those of your intended audience. Just remember—a primary goal is making the text so natural to read, so "invisible," that the reader forgets about individual words and instead enjoys the flow of the reading experience.

6. Does font choice really matter? Yes, actually—it might. Most readers can't express specifically why they do or don't like a book unless it *doesn't feel right*, so it's worth putting some thought into what you want to convey. Do you want an antique look for a classic retelling? See how your book reads in Baskerville. Do you simply prefer ease of readability with a professional appearance? Check out Iowan Old Style or Palatino and see what you think.

When you see how the words fall on these 6"x9" pages, you'll be better able to visualize your finished product. You'll confirm your own reading preferences or cater to those of your intended audience. Just remember—a primary goal is making the text so natural to read, so "invisible," that the reader forgets about individual words and instead enjoys the flow of the reading experience.

7. Does font choice really matter? Yes, actually—it might. Most readers can't express specifically why they do or don't like a book unless it *doesn't feel right*, so it's worth putting some thought into what you want to convey. Do you want an antique look for a classic retelling? See how your book reads in Baskerville. Do you simply prefer ease of readability with a professional appearance? Check out Iowan Old Style or Palatino and see what you think.

When you see how the words fall on these 6"x9" pages, you'll be better able to visualize your finished product. You'll confirm your own reading preferences or cater to those of your intended audience. Just remember—a primary goal is making the text so natural to read, so "invisible," that the reader forgets about individual words and instead enjoys the flow of the reading experience.

IOWAN OLD STYLE

ABCDEFGHIJKLMNOPQRSTUVWXYZ
abcdefghijklmnopqrstuvwxyz

ABCDEFGHIJKLMNOPQRSTUVWXYZ
abcdefghijklmnopqrstuvwxyz

ABCDEFGHIJKLMNOPQRSTUVWXYZ
abcdefghijklmnopqrstuvwxyz

ABCDEFGHIJKLMNOPQRSTUVWXYZ
abcdefghijklmnopqrstuvwxyz

0 1 2 3 4 5 6 7 8 9

IOWAN OLD STYLE
Font sizes 1–7 | Line Spacing: 2

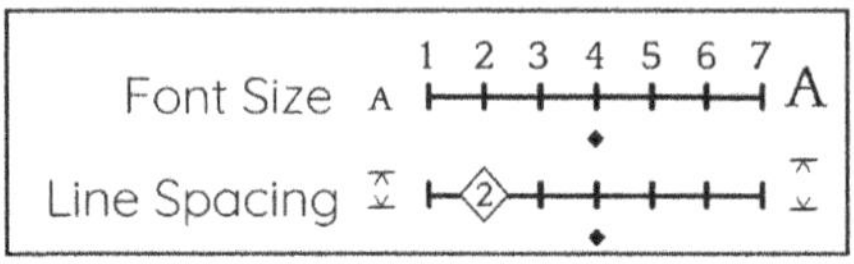

1. **Does font choice really matter?** Yes, actually—it might. Most readers can't express specifically why they do or don't like a book unless it *doesn't feel right*, so it's worth putting some thought into what you want to convey. Do you want an antique look for a classic retelling? See how your book reads in Baskerville. Do you simply prefer ease of readability with a professional appearance? Check out Iowan Old Style or Palatino and see what you think.

When you see how the words fall on these 6"x9" pages, you'll be better able to visualize your finished product. You'll confirm your own reading preferences or cater to those of your intended audience. Just remember—a primary goal is making the text so natural to read, so "invisible," that the reader forgets about individual words and instead enjoys the flow of the reading experience.

2. **Does font choice really matter?** Yes, actually—it might. Most readers can't express specifically why they do or don't like a book unless it *doesn't feel right*, so it's worth putting some thought into what you want to convey. Do you want an antique look for a classic retelling? See how your book reads in Baskerville. Do you simply prefer ease of readability with a professional appearance? Check out Iowan Old Style or Palatino and see what you think.

When you see how the words fall on these 6"x9" pages, you'll be better able to visualize your finished product. You'll confirm your own reading preferences or cater to those of your intended audience. Just remember—a primary goal is making the text so natural to read, so "invisible," that the reader forgets about individual words and instead enjoys the flow of the reading experience.

3. **Does font choice really matter?** Yes, actually—it might. Most readers can't express specifically why they do or don't like a book unless it *doesn't feel right*, so it's worth putting some thought into what you want to convey. Do you want an antique look for a classic retelling? See how your book reads in Baskerville. Do you simply prefer ease of readability with a professional appearance? Check out Iowan Old Style or Palatino and see what you think.

When you see how the words fall on these 6"x9" pages, you'll be better able to visualize your finished product. You'll confirm your own reading preferences or cater to those of your intended audience. Just remember—a primary goal is making the text so natural to read, so "invisible," that the reader forgets about individual words and instead enjoys the flow of the reading experience.

4. **Does font choice really matter?** Yes, actually—it might. Most readers can't express specifically why they do or don't like a book unless it *doesn't feel right,* so it's worth putting some thought into what you want to convey. Do you want an antique look for a classic retelling? See how your book reads in Baskerville. Do you simply prefer ease of readability with a professional appearance? Check out Iowan Old Style or Palatino and see what you think.

When you see how the words fall on these 6"x9" pages, you'll be better able to visualize your finished product. You'll confirm your own reading preferences or cater to those of your intended audience. Just remember—a primary goal is making the text so natural to read, so "invisible," that the reader forgets about individual words and instead enjoys the flow of the reading experience.

5. **Does font choice really matter?** Yes, actually—it might. Most readers can't express specifically why they do or don't like a book unless it *doesn't feel right,* so it's worth putting some thought into what you want to convey. Do you want an antique look for a classic retelling? See how your book reads in Baskerville. Do you simply prefer ease of readability with a professional appearance? Check out Iowan Old Style or Palatino and see what you think.

When you see how the words fall on these 6"x9" pages, you'll be better able to visualize your finished product. You'll confirm your own reading preferences or cater to those of your intended audience. Just remember—a primary goal is making the text so natural to read, so "invisible," that the reader forgets about individual words and instead enjoys the flow of the reading experience.

6. Does font choice really matter? Yes, actually—it might. Most readers can't express specifically why they do or don't like a book unless it *doesn't feel right*, so it's worth putting some thought into what you want to convey. Do you want an antique look for a classic retelling? See how your book reads in Baskerville. Do you simply prefer ease of readability with a professional appearance? Check out Iowan Old Style or Palatino and see what you think.

When you see how the words fall on these 6"x9" pages, you'll be better able to visualize your finished product. You'll confirm your own reading preferences or cater to those of your intended audience. Just remember—a primary goal is making the text so natural to read, so "invisible," that the reader forgets about individual words and instead enjoys the flow of the reading experience.

7. Does font choice really matter? Yes, actually—it might. Most readers can't express specifically why they do or don't like a book unless it *doesn't feel right*, so it's worth putting some thought into what you want to convey. Do you want an antique look for a classic retelling? See how your book reads in Baskerville. Do you simply prefer ease of readability with a professional appearance? Check out Iowan Old Style or Palatino and see what you think.

When you see how the words fall on these 6"x9" pages, you'll be better able to visualize your finished product. You'll confirm your own reading preferences or cater to those of your intended audience. Just remember—a primary goal is making the text so natural to read, so "invisible," that the reader forgets about individual words and instead enjoys the flow of the reading experience.

PALATINO

ABCDEFGHIJKLMNOPQRSTUVWXYZ
abcdefghijklmnopqrstuvwxyz

ABCDEFGHIJKLMNOPQRSTUVWXYZ
abcdefghijklmnopqrstuvwxyz

ABCDEFGHIJKLMNOPQRSTUVWXYZ
abcdefghijklmnopqrstuvwxyz

ABCDEFGHIJKLMNOPQRSTUVWXYZ
abcdefghijklmnopqrstuvwxyz

0 1 2 3 4 5 6 7 8 9

PALATINO
Font sizes 1–7 | Line Spacing: 2

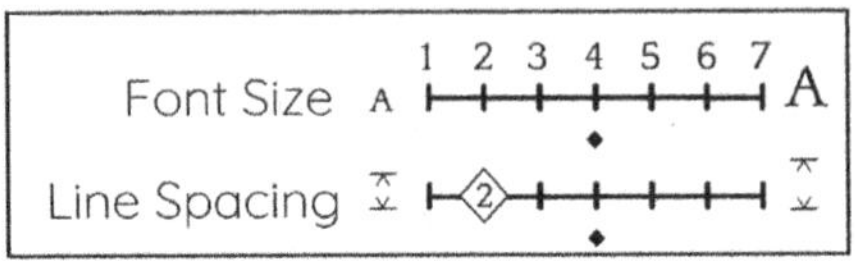

1. Does font choice really matter? Yes, actually—it might. Most readers can't express specifically why they do or don't like a book unless it *doesn't feel right,* so it's worth putting some thought into what you want to convey. Do you want an antique look for a classic retelling? See how your book reads in Baskerville. Do you simply prefer ease of readability with a professional appearance? Check out Iowan Old Style or Palatino and see what you think.

When you see how the words fall on these 6″x9″ pages, you'll be better able to visualize your finished product. You'll confirm your own reading preferences or cater to those of your intended audience. Just remember—a primary goal is making the text so natural to read, so "invisible," that the reader forgets about individual words and instead enjoys the flow of the reading experience.

2. Does font choice really matter? Yes, actually—it might. Most readers can't express specifically why they do or don't like a book unless it *doesn't feel right,* so it's worth putting some thought into what you want to convey. Do you want an antique look for a classic retelling? See how your book reads in Baskerville. Do you simply prefer ease of readability with a professional appearance? Check out Iowan Old Style or Palatino and see what you think.

When you see how the words fall on these 6″x9″ pages, you'll be better able to visualize your finished product. You'll confirm your own reading preferences or cater to those of your intended audience. Just remember—a primary goal is making the text so natural to read, so "invisible," that the reader forgets about individual words and instead enjoys the flow of the reading experience.

3. Does font choice really matter? Yes, actually—it might. Most readers can't express specifically why they do or don't like a book unless it *doesn't feel right,* so it's worth putting some thought into what you want to convey. Do you want an antique look for a classic retelling? See how your book reads in Baskerville. Do you simply prefer ease of readability with a professional appearance? Check out Iowan Old Style or Palatino and see what you think.

When you see how the words fall on these 6"x9" pages, you'll be better able to visualize your finished product. You'll confirm your own reading preferences or cater to those of your intended audience. Just remember—a primary goal is making the text so natural to read, so "invisible," that the reader forgets about individual words and instead enjoys the flow of the reading experience.

4. Does font choice really matter? Yes, actually—it might. Most readers can't express specifically why they do or don't like a book unless it *doesn't feel right,* so it's worth putting some thought into what you want to convey. Do you want an antique look for a classic retelling? See how your book reads in Baskerville. Do you simply prefer ease of readability with a professional appearance? Check out Iowan Old Style or Palatino and see what you think.

When you see how the words fall on these 6"x9" pages, you'll be better able to visualize your finished product. You'll confirm your own reading preferences or cater to those of your intended audience. Just remember—a primary goal is making the text so natural to read, so "invisible," that the reader forgets about individual words and instead enjoys the flow of the reading experience.

5. Does font choice really matter? Yes, actually—it might. Most readers can't express specifically why they do or don't like a book unless it *doesn't feel right,* so it's worth putting some thought into what you want to convey. Do you want an antique look for a classic retelling? See how your book reads in Baskerville. Do you simply prefer ease of readability with a professional appearance? Check out Iowan Old Style or Palatino and see what you think.

When you see how the words fall on these 6"x9" pages, you'll be better able to visualize your finished product. You'll confirm your own reading preferences or cater to those of your intended audience. Just remember—a primary goal is making the text so natural to read, so "invisible," that the reader forgets about individual words and instead enjoys the flow of the reading experience.

6. Does font choice really matter? Yes, actually—it might. Most readers can't express specifically why they do or don't like a book unless it *doesn't feel right,* so it's worth putting some thought into what you want to convey. Do you want an antique look for a classic retelling? See how your book reads in Baskerville. Do you simply prefer ease of readability with a professional appearance? Check out Iowan Old Style or Palatino and see what you think.

When you see how the words fall on these 6"x9" pages, you'll be better able to visualize your finished product. You'll confirm your own reading preferences or cater to those of your intended audience. Just remember—a primary goal is making the text so natural to read, so "invisible," that the reader forgets about individual words and instead enjoys the flow of the reading experience.

7. Does font choice really matter? Yes, actually—it might. Most readers can't express specifically why they do or don't like a book unless it *doesn't feel right,* so it's worth putting some thought into what you want to convey. Do you want an antique look for a classic retelling? See how your book reads in Baskerville. Do you simply prefer ease of readability with a professional appearance? Check out Iowan Old Style or Palatino and see what you think.

When you see how the words fall on these 6"x9" pages, you'll be better able to visualize your finished product. You'll confirm your own reading preferences or cater to those of your intended audience. Just remember—a primary goal is making the text so natural to read, so "invisible," that the reader forgets about individual words and instead enjoys the flow of the reading experience.

TIMES NEW ROMAN

ABCDEFGHIJKLMNOPQRSTUVWXYZ
abcdefghijklmnopqrstuvwxyz

ABCDEFGHIJKLMNOPQRSTUVWXYZ
abcdefghijklmnopqrstuvwxyz

ABCDEFGHIJKLMNOPQRSTUVWXYZ
abcdefghijklmnopqrstuvwxyz

ABCDEFGHIJKLMNOPQRSTUVWXYZ
abcdefghijklmnopqrstuvwxyz

0 1 2 3 4 5 6 7 8 9

TIMES NEW ROMAN

Font sizes 1-7 | Line Spacing: 2

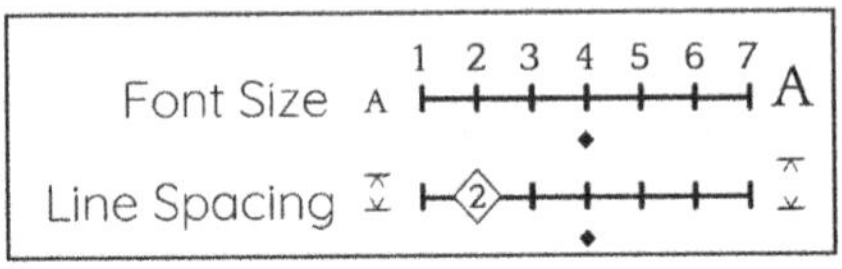

1. Does font choice really matter? Yes, actually—it might. Most readers can't express specifically why they do or don't like a book unless it *doesn't feel right*, so it's worth putting some thought into what you want to convey. Do you want an antique look for a classic retelling? See how your book reads in Baskerville. Do you simply prefer ease of readability with a professional appearance? Check out Iowan Old Style or Palatino and see what you think.

When you see how the words fall on these 6"x9" pages, you'll be better able to visualize your finished product. You'll confirm your own reading preferences or cater to those of your intended audience. Just remember—a primary goal is making the text so natural to read, so "invisible," that the reader forgets about individual words and instead enjoys the flow of the reading experience.

2. Does font choice really matter? Yes, actually—it might. Most readers can't express specifically why they do or don't like a book unless it *doesn't feel right*, so it's worth putting some thought into what you want to convey. Do you want an antique look for a classic retelling? See how your book reads in Baskerville. Do you simply prefer ease of readability with a professional appearance? Check out Iowan Old Style or Palatino and see what you think.

When you see how the words fall on these 6"x9" pages, you'll be better able to visualize your finished product. You'll confirm your own reading preferences or cater to those of your intended audience. Just remember—a primary goal is making the text so natural to read, so "invisible," that the reader forgets about individual words and instead enjoys the flow of the reading experience.

3. Does font choice really matter? Yes, actually—it might. Most readers can't express specifically why they do or don't like a book unless it *doesn't feel right*, so it's worth putting some thought into what you want to convey. Do you want an antique look for a classic retelling? See how your book reads in Baskerville. Do you simply prefer ease of readability with a professional appearance? Check out Iowan Old Style or Palatino and see what you think.

When you see how the words fall on these 6"x9" pages, you'll be better able to visualize your finished product. You'll confirm your own reading preferences or cater to those of your intended audience. Just remember—a primary goal is making the text so natural to read, so "invisible," that the reader forgets about individual words and instead enjoys the flow of the reading experience.

4. Does font choice really matter? Yes, actually—it might. Most readers can't express specifically why they do or don't like a book unless it *doesn't feel right*, so it's worth putting some thought into what you want to convey. Do you want an antique look for a classic retelling? See how your book reads in Baskerville. Do you simply prefer ease of readability with a professional appearance? Check out Iowan Old Style or Palatino and see what you think.

When you see how the words fall on these 6"x9" pages, you'll be better able to visualize your finished product. You'll confirm your own reading preferences or cater to those of your intended audience. Just remember—a primary goal is making the text so natural to read, so "invisible," that the reader forgets about individual words and instead enjoys the flow of the reading experience.

5. Does font choice really matter? Yes, actually—it might. Most readers can't express specifically why they do or don't like a book unless it *doesn't feel right*, so it's worth putting some thought into what you want to convey. Do you want an antique look for a classic retelling? See how your book reads in Baskerville. Do you simply prefer ease of readability with a professional appearance? Check out Iowan Old Style or Palatino and see what you think.

When you see how the words fall on these 6"x9" pages, you'll be better able to visualize your finished product. You'll confirm your own reading preferences or cater to those of your intended audience. Just remember—a primary goal is making the text so natural to read, so "invisible," that the reader forgets about individual words and instead enjoys the flow of the reading experience.

6. Does font choice really matter? Yes, actually—it might. Most readers can't express specifically why they do or don't like a book unless it *doesn't feel right*, so it's worth putting some thought into what you want to convey. Do you want an antique look for a classic retelling? See how your book reads in Baskerville. Do you simply prefer ease of readability with a professional appearance? Check out Iowan Old Style or Palatino and see what you think.

When you see how the words fall on these 6"x9" pages, you'll be better able to visualize your finished product. You'll confirm your own reading preferences or cater to those of your intended audience. Just remember—a primary goal is making the text so natural to read, so "invisible," that the reader forgets about individual words and instead enjoys the flow of the reading experience.

7. Does font choice really matter? Yes, actually—it might. Most readers can't express specifically why they do or don't like a book unless it *doesn't feel right*, so it's worth putting some thought into what you want to convey. Do you want an antique look for a classic retelling? See how your book reads in Baskerville. Do you simply prefer ease of readability with a professional appearance? Check out Iowan Old Style or Palatino and see what you think.

When you see how the words fall on these 6"x9" pages, you'll be better able to visualize your finished product. You'll confirm your own reading preferences or cater to those of your intended audience. Just remember—a primary goal is making the text so natural to read, so "invisible," that the reader forgets about individual words and instead enjoys the flow of the reading experience.

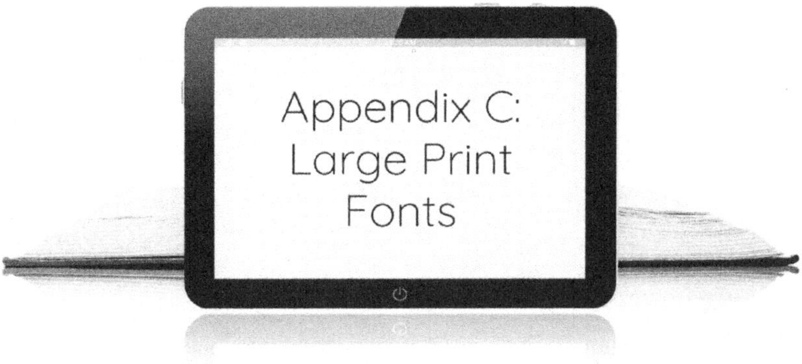

Appendix C: Large Print Fonts

In February 2019, the 180g team answered Vellum users' requests to add large print options for print PDFs. In the File > Print Settings menu, under the More Options button, you'll find the Large Print trim sizes at the very bottom. Selecting one of these sizes automatically switches the font formatting to their larger versions, as you can see from comparing size 7 in Appendix B with size 1 of the same font in Appendix C.

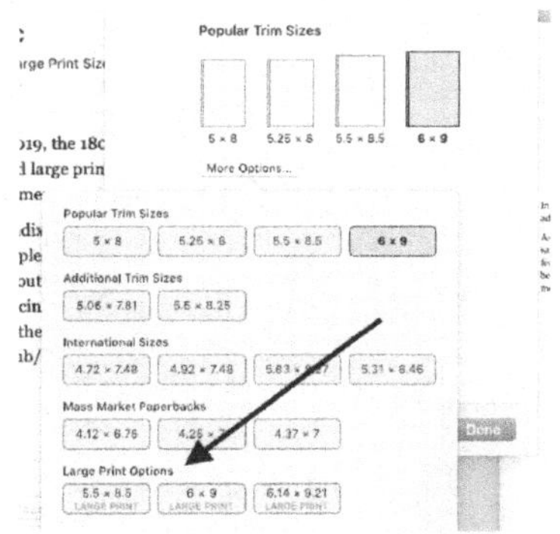

As in Appendix B, each sample in its font grouping is numbered 1 through 7, according to the "Font Size" slider settings in the Print Settings dialog. All samples in this segment are generated at the same "Line Spacing" slider setting (#1—large print text takes up a lot of room, and readers are accustomed to tighter spacing and smaller margins; check your local bookseller to see examples).

And as with earlier samples, these have been inserted at actual size, but you might notice subtle size and spacing differences in your project depending on the text requirements. See the description at help.vellum.pub/print/auto-layout to read more about Vellum's Advanced Layout features .

ADOBE GARAMOND PRO*

ABCDEFGHIJKLMNOPQRSTUVWXYZ
abcdefghijklmnopqrstuvwxyz

ABCDEFGHIJKLMNOPQRSTUVWXYZ
abcdefghijklmnopqrstuvwxyz

ABCDEFGHIJKLMNOPQRSTUVWXYZ
abcdefghijklmnopqrstuvwxyz

ABCDEFGHIJKLMNOPQRSTUVWXYZ
abcdefghijklmnopqrstuvwxyz

0 1 2 3 4 5 6 7 8 9

*Only available if already installed/activated on your computer or server. See "Styles" chapter for more details.

ADOBE GARAMOND PRO
Font sizes 1–7 | Line Spacing: 1

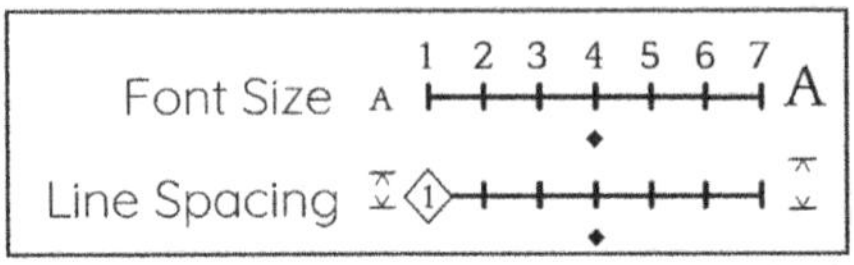

1. Does font choice really matter? While most readers can't express specifically why they do or don't like a book, it's worth putting some thought into what you want them to visualize. Do you want a classic look? See how your book reads in Baskerville. Do you simply prefer ease of readability with a professional appearance? Check out Iowan Old Style or Palatino.

Just remember—a primary goal is making the text so natural to read, so "invisible," that the reader forgets about individual words and enjoys the flow of the reading experience.

2. Does font choice really matter? While most readers can't express specifically why they do or don't like a book, it's worth putting some thought into what you want them to visualize. Do you want a classic look? See how your book reads in Baskerville. Do you simply prefer ease of readability with a professional appearance? Check out Iowan Old Style or Palatino.

Just remember—a primary goal is making the text so natural to read, so "invisible," that the reader forgets about individual words and enjoys the flow of the reading experience.

3. Does font choice really matter? While most readers can't express specifically why they do or don't like a book, it's worth putting some thought into what you want them to visualize. Do you want a classic look? See how your book reads in Baskerville. Do you simply prefer ease of readability with a professional appearance? Check out Iowan Old Style or Palatino.

Just remember—a primary goal is making the text so natural to read, so "invisible," that the reader forgets about individual words and enjoys the flow of the reading experience.

4. Does font choice really matter? While most readers can't express specifically why they do or don't like a book, it's worth putting some thought into what you want them to visualize. Do you want a classic look? See how your book reads in Baskerville. Do you simply prefer ease of readability with a professional appearance? Check out Iowan Old Style or Palatino.

Just remember—a primary goal is making the text so natural to read, so "invisible," that the reader forgets about individual words and enjoys the flow of the reading experience.

5. Does font choice really matter? While most readers can't express specifically why they do or don't like a book, it's worth putting some thought into what you want them to visualize. Do you want a classic look? See how your book reads in Baskerville. Do you simply prefer ease of readability with a professional appearance? Check out Iowan Old Style or Palatino.

Just remember—a primary goal is making the text so natural to read, so "invisible," that the reader forgets about individual words and enjoys the flow of the reading experience.

6. Does font choice really matter? While most readers can't express specifically why they do or don't like a book, it's worth putting some thought into what you want them to visualize. Do you want a classic look? See how your book reads in Baskerville. Do you simply prefer ease of readability with a professional appearance? Check out Iowan Old Style or Palatino.

Just remember—a primary goal is making the text so natural to read, so "invisible," that the reader forgets about individual words and enjoys the flow of the reading experience.

7. Does font choice really matter? While most readers can't express specifically why they do or don't like a book, it's worth putting some thought into what you want them to visualize. Do you want a classic look? See how your book reads in Baskerville. Do you simply prefer ease of readability with a professional appearance? Check out Iowan Old Style or Palatino.

Just remember—a primary goal is making the text so natural to read, so "invisible," that the reader forgets about individual words and enjoys the flow of the reading experience.

ATHELAS

ABCDEFGHIJKLMNOPQRSTUVWXYZ
abcdefghijklmnopqrstuvwxyz

ABCDEFGHIJKLMNOPQRSTUVWXYZ
abcdefghijklmnopqrstuvwxyz

ABCDEFGHIJKLMNOPQRSTUVWXYZ
abcdefghijklmnopqrstuvwxyz

ABCDEFGHIJKLMNOPQRSTUVWXYZ
abcdefghijklmnopqrstuvwxyz

0 1 2 3 4 5 6 7 8 9

ATHELAS

Font sizes 1–7 | Line Spacing: 1

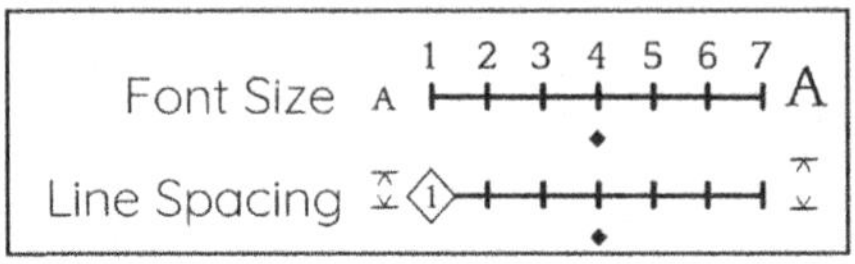

1. Does font choice really matter? While most readers can't express specifically why they do or don't like a book, it's worth putting some thought into what you want them to visualize. Do you want a classic look? See how your book reads in Baskerville. Do you simply prefer ease of readability with a professional appearance? Check out Iowan Old Style or Palatino.

Just remember—a primary goal is making the text so natural to read, so "invisible," that the reader forgets about individual words and enjoys the flow of the reading experience.

2. Does font choice really matter? While most readers can't express specifically why they do or don't like a book, it's worth putting some thought into what you want them to visualize. Do you want a classic look? See how your book reads in Baskerville. Do you simply prefer ease of readability with a professional appearance? Check out Iowan Old Style or Palatino.

Just remember—a primary goal is making the text so natural to read, so "invisible," that the reader forgets about individual words and enjoys the flow of the reading experience.

3. Does font choice really matter? While most readers can't express specifically why they do or don't like a book, it's worth putting some thought into what you want them to visualize. Do you want a

classic look? See how your book reads in Baskerville. Do you simply prefer ease of readability with a professional appearance? Check out Iowan Old Style or Palatino.

Just remember—a primary goal is making the text so natural to read, so "invisible," that the reader forgets about individual words and enjoys the flow of the reading experience.

4. **Does font choice really matter?** While most readers can't express specifically why they do or don't like a book, it's worth putting some thought into what you want them to visualize. Do you want a classic look? See how your book reads in Baskerville. Do you simply prefer ease of readability with a professional appearance? Check out Iowan Old Style or Palatino.

Just remember—a primary goal is making the text so natural to read, so "invisible," that the reader forgets about individual words and enjoys the flow of the reading experience.

5. **Does font choice really matter?** While most readers can't express specifically why they do or don't like a book, it's worth putting some thought into what you want them to visualize. Do you want a classic look? See how your book reads in Baskerville. Do you simply prefer ease of readability with a professional appearance? Check out Iowan Old Style or Palatino.

Just remember—a primary goal is making the text so natural to read, so "invisible," that the

reader forgets about individual words and enjoys the flow of the reading experience.

6. Does font choice really matter? While most readers can't express specifically why they do or don't like a book, it's worth putting some thought into what you want them to visualize. Do you want a classic look? See how your book reads in Baskerville. Do you simply prefer ease of readability with a professional appearance? Check out Iowan Old Style or Palatino.

Just remember—a primary goal is making the text so natural to read, so "invisible," that the reader forgets about individual words and enjoys the flow of the reading experience.

7. Does font choice really matter? While most readers can't express specifically why they do or don't like a book, it's worth putting some thought into what you want them to visualize. Do you want a classic look? See how your book reads in Baskerville. Do you simply prefer ease of readability with a professional appearance? Check out Iowan Old Style or Palatino.

Just remember—a primary goal is making the text so natural to read, so "invisible," that the reader forgets about individual words and enjoys the flow of the reading experience.

BASKERVILLE

ABCDEFGHIJKLMNOPQRSTUVWXYZ
abcdefghijklmnopqrstuvwxyz

ABCDEFGHIJKLMNOPQRSTUVWXYZ
abcdefghijklmnopqrstuvwxyz

ABCDEFGHIJKLMNOPQRSTUVWXYZ
abcdefghijklmnopqrstuvwxyz

ABCDEFGHIJKLMNOPQRSTUVWXYZ
abcdefghijklmnopqrstuvwxyz

0 1 2 3 4 5 6 7 8 9

BASKERVILLE
Font sizes 1–7 | Line Spacing: 1

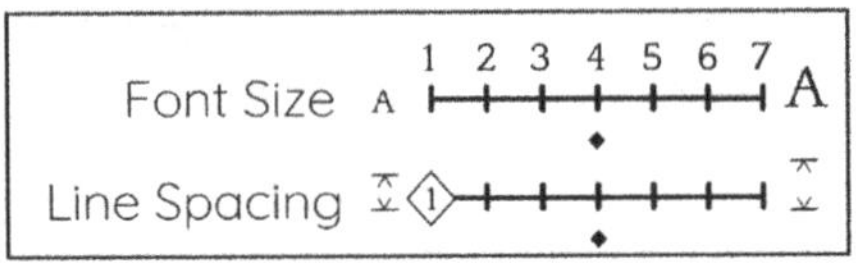

1. Does font choice really matter? While most readers can't express specifically why they do or don't like a book, it's worth putting some thought into what you want them to visualize. Do you want a classic look? See how your book reads in Baskerville. Do you simply prefer ease of readability with a professional appearance? Check out Iowan Old Style or Palatino.

Just remember—a primary goal is making the text so natural to read, so "invisible," that the reader forgets about individual words and enjoys the flow of the reading experience.

2. Does font choice really matter? While most readers can't express specifically why they do or don't like a book, it's worth putting some thought into what you want them to visualize. Do you want a classic look? See how your book reads in Baskerville. Do you simply prefer ease of readability with a professional appearance? Check out Iowan Old Style or Palatino.

Just remember—a primary goal is making the text so natural to read, so "invisible," that the reader forgets about individual words and enjoys the flow of the reading experience.

3. Does font choice really matter? While most readers can't express specifically why they do or don't like a book, it's worth putting some thought into what you want them to visualize. Do you want

a classic look? See how your book reads in Baskerville. Do you simply prefer ease of readability with a professional appearance? Check out Iowan Old Style or Palatino.

Just remember—a primary goal is making the text so natural to read, so “invisible,” that the reader forgets about individual words and enjoys the flow of the reading experience.

4. Does font choice really matter? While most readers can’t express specifically why they do or don’t like a book, it’s worth putting some thought into what you want them to visualize. Do you want a classic look? See how your book reads in Baskerville. Do you simply prefer ease of readability with a professional appearance? Check out Iowan Old Style or Palatino.

Just remember—a primary goal is making the text so natural to read, so “invisible,” that the reader forgets about individual words and enjoys the flow of the reading experience.

5. Does font choice really matter? While most readers can’t express specifically why they do or don’t like a book, it’s worth putting some thought into what you want them to visualize. Do you want a classic look? See how your book reads in Baskerville. Do you simply prefer ease of readability with a professional appearance? Check out Iowan Old Style or Palatino.

Just remember—a primary goal is making the text so natural to read, so “invisible,” that the

reader forgets about individual words and enjoys the flow of the reading experience.

6. Does font choice really matter? While most readers can't express specifically why they do or don't like a book, it's worth putting some thought into what you want them to visualize. Do you want a classic look? See how your book reads in Baskerville. Do you simply prefer ease of readability with a professional appearance? Check out Iowan Old Style or Palatino.

Just remember—a primary goal is making the text so natural to read, so "invisible," that the reader forgets about individual words and enjoys the flow of the reading experience.

7. Does font choice really matter? While most readers can't express specifically why they do or don't like a book, it's worth putting some thought into what you want them to visualize. Do you want a classic look? See how your book reads in Baskerville. Do you simply prefer ease of readability with a professional appearance? Check out Iowan Old Style or Palatino.

Just remember—a primary goal is making the text so natural to read, so "invisible," that the reader forgets about individual words and enjoys the flow of the reading experience.

COCHIN

ABCDEFGHIJKLMNOPQRSTUVWXYZ
abcdefghijklmnopqrstuvwxyz

ABCDEFGHIJKLMNOPQRSTUVWXYZ
abcdefghijklmnopqrstuvwxyz

ABCDEFGHIJKLMNOPQRSTUVWXYZ
abcdefghijklmnopqrstuvwxyz

ABCDEFGHIJKLMNOPQRSTUVWXYZ
abcdefghijklmnopqrstuvwxyz

0 1 2 3 4 5 6 7 8 9

COCHIN

Font sizes 1–7 | Line Spacing: 1

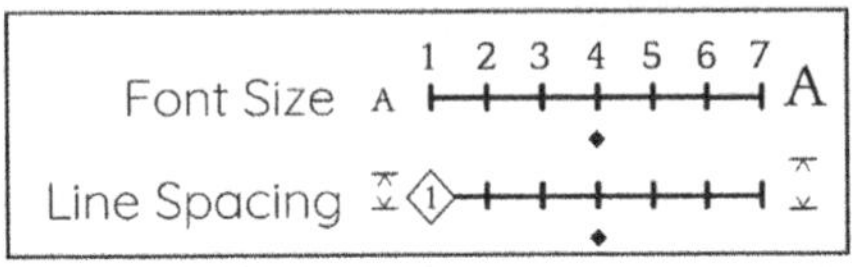

1. Does font choice really matter? While most readers can't express specifically why they do or don't like a book, it's worth putting some thought into what you want them to visualize. Do you want a classic look? See how your book reads in Baskerville. Do you simply prefer ease of readability with a professional appearance? Check out Iowan Old Style or Palatino.

Just remember—a primary goal is making the text so natural to read, so "invisible," that the reader forgets about individual words and enjoys the flow of the reading experience.

2. Does font choice really matter? While most readers can't express specifically why they do or don't like a book, it's worth putting some thought into what you want them to visualize. Do you want a classic look? See how your book reads in Baskerville. Do you simply prefer ease of readability with a professional appearance? Check out Iowan Old Style or Palatino.

Just remember—a primary goal is making the text so natural to read, so "invisible," that the reader forgets about individual words and enjoys the flow of the reading experience.

3. Does font choice really matter? While most readers can't express specifically why they do or don't like a book, it's worth putting some thought into what you want them to visualize. Do you want a classic look? See how your book reads in Baskerville. Do you simply prefer ease of readability with a professional appearance? Check out Iowan Old Style or Palatino.

Just remember—a primary goal is making the text so natural to read, so "invisible," that the reader forgets about individual words and enjoys the flow of the reading experience.

4. Does font choice really matter? While most readers can't express specifically why they do or don't like a book, it's worth putting some thought into what you want them to visualize. Do you want a classic look? See how your book reads in Baskerville. Do you simply prefer ease of readability with a professional appearance? Check out Iowan Old Style or Palatino.

Just remember—a primary goal is making the text so natural to read, so "invisible," that the reader forgets about individual words and enjoys the flow of the reading experience.

5. Does font choice really matter? While most readers can't express specifically why they do or don't like a book, it's worth putting some thought into what you want them to visualize. Do you want a classic look? See how your book reads in Baskerville. Do you simply prefer ease of readability with a professional appearance? Check out Iowan Old Style or Palatino.

Just remember—a primary goal is making the text so natural to read, so "invisible," that the

reader forgets about individual words and enjoys the flow of the reading experience.

6. Does font choice really matter? While most readers can't express specifically why they do or don't like a book, it's worth putting some thought into what you want them to visualize. Do you want a classic look? See how your book reads in Baskerville. Do you simply prefer ease of readability with a professional appearance? Check out Iowan Old Style or Palatino.

Just remember—a primary goal is making the text so natural to read, so "invisible," that the reader forgets about individual words and enjoys the flow of the reading experience.

7. Does font choice really matter? While most readers can't express specifically why they do or don't like a book, it's worth putting some thought into what you want them to visualize. Do you want a classic look? See how your book reads in Baskerville. Do you simply prefer ease of readability with a professional appearance? Check out Iowan Old Style or Palatino.

Just remember—a primary goal is making the text so natural to read, so "invisible," that the reader forgets about individual words and enjoys the flow of the reading experience.

CRIMSON TEXT

ABCDEFGHIJKLMNOPQRSTUVWXYZ
abcdefghijklmnopqrstuvwxyz

ABCDEFGHIJKLMNOPQRSTUVWXYZ
abcdefghijklmnopqrstuvwxyz

ABCDEFGHIJKLMNOPQRSTUVWXYZ
abcdefghijklmnopqrstuvwxyz

ABCDEFGHIJKLMNOPQRSTUVWXYZ
abcdefghijklmnopqrstuvwxyz

0 1 2 3 4 5 6 7 8 9

CRIMSON TEXT
Font sizes 1–7 | Line Spacing: 1

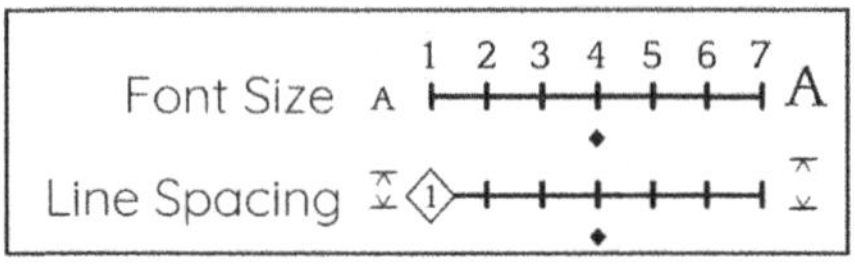

1. Does font choice really matter? While most readers can't express specifically why they do or don't like a book, it's worth putting some thought into what you want them to visualize. Do you want a classic look? See how your book reads in Baskerville. Do you simply prefer ease of readability with a professional appearance? Check out Iowan Old Style or Palatino.

Just remember—a primary goal is making the text so natural to read, so "invisible," that the reader forgets about individual words and enjoys the flow of the reading experience.

2. Does font choice really matter? While most readers can't express specifically why they do or don't like a book, it's worth putting some thought into what you want them to visualize. Do you want a classic look? See how your book reads in Baskerville. Do you simply prefer ease of readability with a professional appearance? Check out Iowan Old Style or Palatino.

Just remember—a primary goal is making the text so natural to read, so "invisible," that the reader forgets about individual words and enjoys the flow of the reading experience.

3. Does font choice really matter? While most readers can't express specifically why they do or don't like a book, it's worth putting some thought into what you want them to visualize. Do you want

a classic look? See how your book reads in Baskerville. Do you simply prefer ease of readability with a professional appearance? Check out Iowan Old Style or Palatino.

Just remember—a primary goal is making the text so natural to read, so "invisible," that the reader forgets about individual words and enjoys the flow of the reading experience.

4. **Does font choice really matter?** While most readers can't express specifically why they do or don't like a book, it's worth putting some thought into what you want them to visualize. Do you want a classic look? See how your book reads in Baskerville. Do you simply prefer ease of readability with a professional appearance? Check out Iowan Old Style or Palatino.

Just remember—a primary goal is making the text so natural to read, so "invisible," that the reader forgets about individual words and enjoys the flow of the reading experience.

5. **Does font choice really matter?** While most readers can't express specifically why they do or don't like a book, it's worth putting some thought into what you want them to visualize. Do you want a classic look? See how your book reads in Baskerville. Do you simply prefer ease of readability with a professional appearance? Check out Iowan Old Style or Palatino.

Just remember—a primary goal is making the text so natural to read, so "invisible," that the reader forgets about individual words and enjoys the flow of the reading experience.

6. **Does font choice really matter?** While most readers can't express specifically why they do or don't like a book, it's worth putting some thought into what you want them to visualize. Do you want a classic look? See how your book reads in Baskerville. Do you simply prefer ease of readability with a professional appearance? Check out Iowan Old Style or Palatino.

Just remember—a primary goal is making the text so natural to read, so "invisible," that the reader forgets about individual words and enjoys the flow of the reading experience.

7. **Does font choice really matter?** While most readers can't express specifically why they do or don't like a book, it's worth putting some thought into what you want them to visualize. Do you want a classic look? See how your book reads in Baskerville. Do you simply prefer ease of readability with a professional appearance? Check out Iowan Old Style or Palatino.

Just remember—a primary goal is making the text so natural to read, so "invisible," that the reader forgets about individual words and enjoys the flow of the reading experience.

FANWOOD

ABCDEFGHIJKLMNOPQRSTUVWXYZ
abcdefghijklmnopqrstuvwxyz

ABCDEFGHIJKLMNOPQRSTUVWXYZ
abcdefghijklmnopqrstuvwxyz

ABCDEFGHIJKLMNOPQRSTUVWXYZ
abcdefghijklmnopqrstuvwxyz

ABCDEFGHIJKLMNOPQRSTUVWXYZ
abcdefghijklmnopqrstuvwxyz

0 1 2 3 4 5 6 7 8 9

FANWOOD

Font sizes 1–7 | Line Spacing: 1

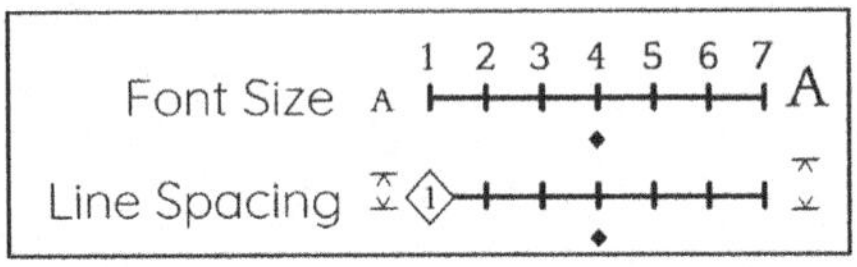

1. Does font choice really matter? While most readers can't express specifically why they do or don't like a book, it's worth putting some thought into what you want them to visualize. Do you want a classic look? See how your book reads in Baskerville. Do you simply prefer ease of readability with a professional appearance? Check out Iowan Old Style or Palatino.

Just remember—a primary goal is making the text so natural to read, so "invisible," that the reader forgets about individual words and enjoys the flow of the reading experience.

2. Does font choice really matter? While most readers can't express specifically why they do or don't like a book, it's worth putting some thought into what you want them to visualize. Do you want a classic look? See how your book reads in Baskerville. Do you simply prefer ease of readability with a professional appearance? Check out Iowan Old Style or Palatino.

Just remember—a primary goal is making the text so natural to read, so "invisible," that the reader forgets about individual words and enjoys the flow of the reading experience.

3. Does font choice really matter? While most readers can't express specifically why they do or don't like a book, it's worth putting some thought into what you want them to visualize. Do you want a

classic look? See how your book reads in Baskerville. Do you simply prefer ease of readability with a professional appearance? Check out Iowan Old Style or Palatino.

Just remember—a primary goal is making the text so natural to read, so "invisible," that the reader forgets about individual words and enjoys the flow of the reading experience.

4. Does font choice really matter? While most readers can't express specifically why they do or don't like a book, it's worth putting some thought into what you want them to visualize. Do you want a classic look? See how your book reads in Baskerville. Do you simply prefer ease of readability with a professional appearance? Check out Iowan Old Style or Palatino.

Just remember—a primary goal is making the text so natural to read, so "invisible," that the reader forgets about individual words and enjoys the flow of the reading experience.

5. Does font choice really matter? While most readers can't express specifically why they do or don't like a book, it's worth putting some thought into what you want them to visualize. Do you want a classic look? See how your book reads in Baskerville. Do you simply prefer ease of readability with a professional appearance? Check out Iowan Old Style or Palatino.

Just remember—a primary goal is making the text so natural to read, so "invisible," that the

reader forgets about individual words and enjoys the flow of the reading experience.

6. Does font choice really matter? While most readers can't express specifically why they do or don't like a book, it's worth putting some thought into what you want them to visualize. Do you want a classic look? See how your book reads in Baskerville. Do you simply prefer ease of readability with a professional appearance? Check out Iowan Old Style or Palatino.

Just remember—a primary goal is making the text so natural to read, so "invisible," that the reader forgets about individual words and enjoys the flow of the reading experience.

7. Does font choice really matter? While most readers can't express specifically why they do or don't like a book, it's worth putting some thought into what you want them to visualize. Do you want a classic look? See how your book reads in Baskerville. Do you simply prefer ease of readability with a professional appearance? Check out Iowan Old Style or Palatino.

Just remember—a primary goal is making the text so natural to read, so "invisible," that the reader forgets about individual words and enjoys the flow of the reading experience.

HOEFLER TEXT

ABCDEFGHIJKLMNOPQRSTUVWXYZ
abcdefghijklmnopqrstuvwxyz

ABCDEFGHIJKLMNOPQRSTUVWXYZ
abcdefghijklmnopqrstuvwxyz

ABCDEFGHIJKLMNOPQRSTUVWXYZ
abcdefghijklmnopqrstuvwxyz

ABCDEFGHIJKLMNOPQRSTUVWXYZ
abcdefghijklmnopqrstuvwxyz

0 1 2 3 4 5 6 7 8 9

HOEFLER TEXT

Font sizes 1-7 | Line Spacing: 1

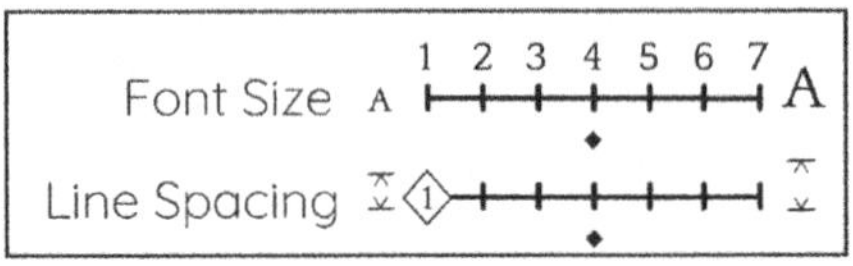

1. Does font choice really matter? While most readers can't express specifically why they do or don't like a book, it's worth putting some thought into what you want them to visualize. Do you want a classic look? See how your book reads in Baskerville. Do you simply prefer ease of readability with a professional appearance? Check out Iowan Old Style or Palatino.

Just remember—a primary goal is making the text so natural to read, so "invisible," that the reader forgets about individual words and enjoys the flow of the reading experience.

2. Does font choice really matter? While most readers can't express specifically why they do or don't like a book, it's worth putting some thought into what you want them to visualize. Do you want a classic look? See how your book reads in Baskerville. Do you simply prefer ease of readability with a professional appearance? Check out Iowan Old Style or Palatino.

Just remember—a primary goal is making the text so natural to read, so "invisible," that the reader forgets about individual words and enjoys the flow of the reading experience.

3. Does font choice really matter? While most readers can't express specifically why they do or don't like a book, it's worth putting some thought into what you want them to visualize. Do you want a

classic look? See how your book reads in Baskerville. Do you simply prefer ease of readability with a professional appearance? Check out Iowan Old Style or Palatino.

Just remember—a primary goal is making the text so natural to read, so "invisible," that the reader forgets about individual words and enjoys the flow of the reading experience.

4. **Does font choice really matter?** While most readers can't express specifically why they do or don't like a book, it's worth putting some thought into what you want them to visualize. Do you want a classic look? See how your book reads in Baskerville. Do you simply prefer ease of readability with a professional appearance? Check out Iowan Old Style or Palatino.

Just remember—a primary goal is making the text so natural to read, so "invisible," that the reader forgets about individual words and enjoys the flow of the reading experience.

5. **Does font choice really matter?** While most readers can't express specifically why they do or don't like a book, it's worth putting some thought into what you want them to visualize. Do you want a classic look? See how your book reads in Baskerville. Do you simply prefer ease of readability with a professional appearance? Check out Iowan Old Style or Palatino.

Just remember—a primary goal is making the text so natural to read, so "invisible," that the

reader forgets about individual words and enjoys the flow of the reading experience.

6. Does font choice really matter? While most readers can't express specifically why they do or don't like a book, it's worth putting some thought into what you want them to visualize. Do you want a classic look? See how your book reads in Baskerville. Do you simply prefer ease of readability with a professional appearance? Check out Iowan Old Style or Palatino.

Just remember—a primary goal is making the text so natural to read, so "invisible," that the reader forgets about individual words and enjoys the flow of the reading experience.

7. Does font choice really matter? While most readers can't express specifically why they do or don't like a book, it's worth putting some thought into what you want them to visualize. Do you want a classic look? See how your book reads in Baskerville. Do you simply prefer ease of readability with a professional appearance? Check out Iowan Old Style or Palatino.

Just remember—a primary goal is making the text so natural to read, so "invisible," that the reader forgets about individual words and enjoys the flow of the reading experience.

IOWAN OLD STYLE

ABCDEFGHIJKLMNOPQRSTUVWXYZ
abcdefghijklmnopqrstuvwxyz

ABCDEFGHIJKLMNOPQRSTUVWXYZ
abcdefghijklmnopqrstuvwxyz

ABCDEFGHIJKLMNOPQRSTUVWXYZ
abcdefghijklmnopqrstuvwxyz

ABCDEFGHIJKLMNOPQRSTUVWXYZ
abcdefghijklmnopqrstuvwxyz

0 1 2 3 4 5 6 7 8 9

IOWAN OLD STYLE
Font sizes 1–7 | Line Spacing: 1

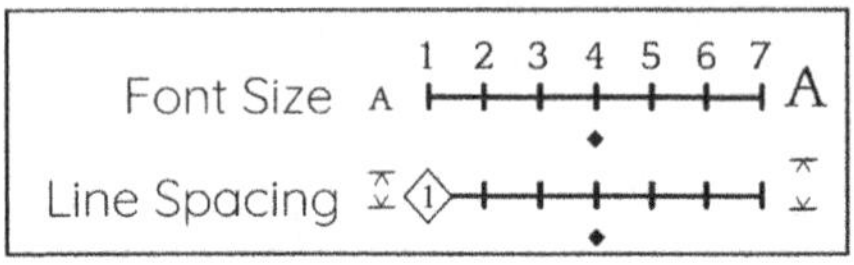

1. **Does font choice really matter?** While most readers can't express specifically why they do or don't like a book, it's worth putting some thought into what you want them to visualize. Do you want a classic look? See how your book reads in Baskerville. Do you simply prefer ease of readability with a professional appearance? Check out Iowan Old Style or Palatino.

Just remember—a primary goal is making the text so natural to read, so "invisible," that the reader forgets about individual words and enjoys the flow of the reading experience.

2. **Does font choice really matter?** While most readers can't express specifically why they do or don't like a book, it's worth putting some thought into what you want them to visualize. Do you want a classic look? See how your book reads in Baskerville. Do you simply prefer ease of readability with a professional appearance? Check out Iowan Old Style or Palatino.

Just remember—a primary goal is making the text so natural to read, so "invisible," that the reader forgets about individual words and enjoys the flow of the reading experience.

3. **Does font choice really matter?** While most readers can't express specifically why they do or don't like a book, it's worth putting some thought into what you want them to visualize. Do you want

a classic look? See how your book reads in Baskerville. Do you simply prefer ease of readability with a professional appearance? Check out Iowan Old Style or Palatino.

Just remember—a primary goal is making the text so natural to read, so “invisible,” that the reader forgets about individual words and enjoys the flow of the reading experience.

4. **Does font choice really matter?** While most readers can’t express specifically why they do or don’t like a book, it’s worth putting some thought into what you want them to visualize. Do you want a classic look? See how your book reads in Baskerville. Do you simply prefer ease of readability with a professional appearance? Check out Iowan Old Style or Palatino.

Just remember—a primary goal is making the text so natural to read, so “invisible,” that the reader forgets about individual words and enjoys the flow of the reading experience.

5. **Does font choice really matter?** While most readers can’t express specifically why they do or don’t like a book, it’s worth putting some thought into what you want them to visualize. Do you want a classic look? See how your book reads in Baskerville. Do you simply prefer ease of readability with a professional appearance? Check out Iowan Old Style or Palatino.

Just remember—a primary goal is making the text so natural to read, so “invisible,” that the reader forgets about individual words and enjoys the flow of the reading experience.

6. Does font choice really matter? While most readers can't express specifically why they do or don't like a book, it's worth putting some thought into what you want them to visualize. Do you want a classic look? See how your book reads in Baskerville. Do you simply prefer ease of readability with a professional appearance? Check out Iowan Old Style or Palatino.

Just remember—a primary goal is making the text so natural to read, so "invisible," that the reader forgets about individual words and enjoys the flow of the reading experience.

7. Does font choice really matter? While most readers can't express specifically why they do or don't like a book, it's worth putting some thought into what you want them to visualize. Do you want a classic look? See how your book reads in Baskerville. Do you simply prefer ease of readability with a professional appearance? Check out Iowan Old Style or Palatino.

Just remember—a primary goal is making the text so natural to read, so "invisible," that the reader forgets about individual words and enjoys the flow of the reading experience.

PALATINO

ABCDEFGHIJKLMNOPQRSTUVWXYZ
abcdefghijklmnopqrstuvwxyz

ABCDEFGHIJKLMNOPQRSTUVWXYZ
abcdefghijklmnopqrstuvwxyz

ABCDEFGHIJKLMNOPQRSTUVWXYZ
abcdefghijklmnopqrstuvwxyz

ABCDEFGHIJKLMNOPQRSTUVWXYZ
abcdefghijklmnopqrstuvwxyz

0 1 2 3 4 5 6 7 8 9

PALATINO

Font sizes 1–7 | Line Spacing: 1

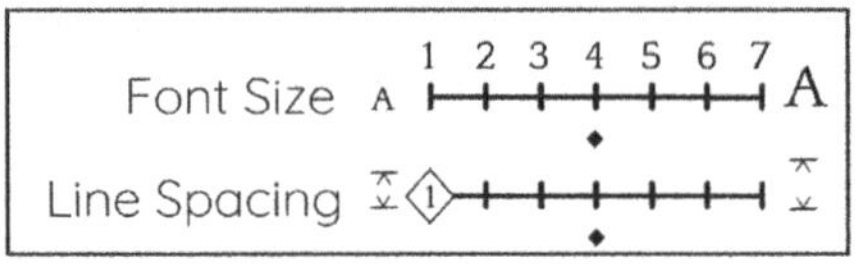

1. Does font choice really matter? While most readers can't express specifically why they do or don't like a book, it's worth putting some thought into what you want them to visualize. Do you want a classic look? See how your book reads in Baskerville. Do you simply prefer ease of readability with a professional appearance? Check out Iowan Old Style or Palatino.

Just remember—a primary goal is making the text so natural to read, so "invisible," that the reader forgets about individual words and enjoys the flow of the reading experience.

2. Does font choice really matter? While most readers can't express specifically why they do or don't like a book, it's worth putting some thought into what you want them to visualize. Do you want a classic look? See how your book reads in Baskerville. Do you simply prefer ease of readability with a professional appearance? Check out Iowan Old Style or Palatino.

Just remember—a primary goal is making the text so natural to read, so "invisible," that the reader forgets about individual words and enjoys the flow of the reading experience.

3. Does font choice really matter? While most readers can't express specifically why they do or don't like a book, it's worth putting some thought into what you want them to visualize. Do you want

a classic look? See how your book reads in Baskerville. Do you simply prefer ease of readability with a professional appearance? Check out Iowan Old Style or Palatino.

Just remember—a primary goal is making the text so natural to read, so "invisible," that the reader forgets about individual words and enjoys the flow of the reading experience.

4. Does font choice really matter? While most readers can't express specifically why they do or don't like a book, it's worth putting some thought into what you want them to visualize. Do you want a classic look? See how your book reads in Baskerville. Do you simply prefer ease of readability with a professional appearance? Check out Iowan Old Style or Palatino.

Just remember—a primary goal is making the text so natural to read, so "invisible," that the reader forgets about individual words and enjoys the flow of the reading experience.

5. Does font choice really matter? While most readers can't express specifically why they do or don't like a book, it's worth putting some thought into what you want them to visualize. Do you want a classic look? See how your book reads in Baskerville. Do you simply prefer ease of readability with a professional appearance? Check out Iowan Old Style or Palatino.

Just remember—a primary goal is making the text so natural to read, so "invisible," that the reader forgets about individual words and enjoys the flow of the reading experience.

6. Does font choice really matter? While most readers can't express specifically why they do or don't like a book, it's worth putting some thought into what you want them to visualize. Do you want a classic look? See how your book reads in Baskerville. Do you simply prefer ease of readability with a professional appearance? Check out Iowan Old Style or Palatino.

Just remember—a primary goal is making the text so natural to read, so "invisible," that the reader forgets about individual words and enjoys the flow of the reading experience.

7. Does font choice really matter? While most readers can't express specifically why they do or don't like a book, it's worth putting some thought into what you want them to visualize. Do you want a classic look? See how your book reads in Baskerville. Do you simply prefer ease of readability with a professional appearance? Check out Iowan Old Style or Palatino.

Just remember—a primary goal is making the text so natural to read, so "invisible," that the reader forgets about individual words and enjoys the flow of the reading experience.

TIMES NEW ROMAN

ABCDEFGHIJKLMNOPQRSTUVWXYZ
abcdefghijklmnopqrstuvwxyz

ABCDEFGHIJKLMNOPQRSTUVWXYZ
abcdefghijklmnopqrstuvwxyz

ABCDEFGHIJKLMNOPQRSTUVWXYZ
abcdefghijklmnopqrstuvwxyz

ABCDEFGHIJKLMNOPQRSTUVWXYZ
abcdefghijklmnopqrstuvwxyz

0 1 2 3 4 5 6 7 8 9

TIMES NEW ROMAN

Font sizes 1–7 | Line Spacing: 1

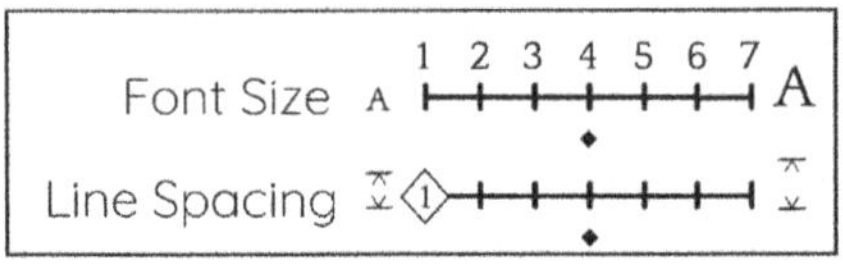

1. Does font choice really matter? While most readers can't express specifically why they do or don't like a book, it's worth putting some thought into what you want them to visualize. Do you want a classic look? See how your book reads in Baskerville. Do you simply prefer ease of readability with a professional appearance? Check out Iowan Old Style or Palatino.

Just remember—a primary goal is making the text so natural to read, so "invisible," that the reader forgets about individual words and enjoys the flow of the reading experience.

2. Does font choice really matter? While most readers can't express specifically why they do or don't like a book, it's worth putting some thought into what you want them to visualize. Do you want a classic look? See how your book reads in Baskerville. Do you simply prefer ease of readability with a professional appearance? Check out Iowan Old Style or Palatino.

Just remember—a primary goal is making the text so natural to read, so "invisible," that the reader forgets about individual words and enjoys the flow of the reading experience.

3. Does font choice really matter? While most readers can't express specifically why they do or don't like a book, it's worth putting some thought into what you want them to visualize. Do you want a

classic look? See how your book reads in Baskerville. Do you simply prefer ease of readability with a professional appearance? Check out Iowan Old Style or Palatino.

Just remember—a primary goal is making the text so natural to read, so "invisible," that the reader forgets about individual words and enjoys the flow of the reading experience.

4. Does font choice really matter? While most readers can't express specifically why they do or don't like a book, it's worth putting some thought into what you want them to visualize. Do you want a classic look? See how your book reads in Baskerville. Do you simply prefer ease of readability with a professional appearance? Check out Iowan Old Style or Palatino.

Just remember—a primary goal is making the text so natural to read, so "invisible," that the reader forgets about individual words and enjoys the flow of the reading experience.

5. Does font choice really matter? While most readers can't express specifically why they do or don't like a book, it's worth putting some thought into what you want them to visualize. Do you want a classic look? See how your book reads in Baskerville. Do you simply prefer ease of readability with a professional appearance? Check out Iowan Old Style or Palatino.

Just remember—a primary goal is making the text so natural to read, so "invisible," that the

reader forgets about individual words and enjoys the flow of the reading experience.

6. Does font choice really matter? While most readers can't express specifically why they do or don't like a book, it's worth putting some thought into what you want them to visualize. Do you want a classic look? See how your book reads in Baskerville. Do you simply prefer ease of readability with a professional appearance? Check out Iowan Old Style or Palatino.

Just remember—a primary goal is making the text so natural to read, so "invisible," that the reader forgets about individual words and enjoys the flow of the reading experience.

7. Does font choice really matter? While most readers can't express specifically why they do or don't like a book, it's worth putting some thought into what you want them to visualize. Do you want a classic look? See how your book reads in Baskerville. Do you simply prefer ease of readability with a professional appearance? Check out Iowan Old Style or Palatino.

Just remember—a primary goal is making the text so natural to read, so "invisible," that the reader forgets about individual words and enjoys the flow of the reading experience.

ACKNOWLEDGMENTS

Creator of All, thank you for discovery. It is one of the greatest joys of my life.

My family, I love you. Since this book began its journey while we were in the middle of moving boxes and DO NOT PACK THIS PILE signs, I thank Mike, the love of my life, and our incredible juveniles, Marrin, Jacob, and Evan. For patiently walking around me as I brainstormed sample text and font weirdness, for answering the door or letting the dog out when my lap was covered with notes, for putting up with a poorly planned supper *again* because my brain was toast . . . For these and so much more, I'm eternally thankful that God blessed me with each of you!

Rogena Mitchell-Jones of RMJ Editing & Manuscript Service—you have been a lifesaver on this project. It's a whole different ballgame on the writing side of the process, and you made it easy for me to hand off! Thank you for being my fellow editor *and* fellow Vellum devotee. And to all our mutual online editor connections . . . You're the best! #Fam #YouKnowWhoYouAre #EditorsNeedEditorsToo

To my editing coach, Jennifer Harshman, and my fellow Shire members, thank you for keeping me accountable to finish this! I'd probably still be revising if it weren't for your encouragement and accountability. And the mastermind folks in the 48 Days Eagles community—you've stretched me to think bigger than I ever dreamed. Thanks to you, too!

JacQueline Roe at The Quirky Quill—your keyword research and social media expertise is nothing short of magical to my brain. Thanks for helping me with this end-of-process craziness!

Todd Damotte—I love this cover! Thanks so much for working with me and bringing your skills to the table. I look forward to future collaborations.

James Woosley at Free Agent Press—I appreciate your expertise and kind help and not thinking my questions are dumb.

I'm amazed by the helpful people I've met in the formatting/book design world, with special mention of the InDesign Secrets and Vellum Users groups on Facebook. For all the questions you've answered for me . . . so many thanks. And what a fun challenge to create a book using both Vellum and InDesign! (EDIT: After I completed the first draft of this book, an opportunity arose to be an admin in the Vellum Users community—and while this IS my "first rodeo" in the Facebook admin world, I couldn't have picked a better group to start with! Thanks to you all.)

For the wonderful writers who have trusted me to edit, proofread, and format your own labor of love—you are MY heroes! Thanks for encouraging me to write my own *real* book.

Last but not least, thanks to The Brads and the 180g team for creating an app that's accessible to everyday users and solidly coded for digital publishing requirements. For your input as I emailed you sections to make sure I represented you and Vellum correctly, as well as for your thoughts and permission on the cover design, I can't thank you enough. Vellum has paid for itself so many times over, and I'm thrilled to point others your way and smooth the path toward many more beautiful books. Keep up the good work, 180g!

ABOUT THE AUTHOR

Jody E. Skinner is a punctuation nerd and spelling aficionado who loves to collaborate, create, and learn, learn, learn. As a military spouse and mom of two teens and a tween, she enjoys the adventure as their family moves around the world.

Jody enjoys the book design universe and solidly believes her team of trusted colleagues—editors, proofreaders, designers, and writers—is THE most valuable commodity she's collected since she officially hung out her proofreader shingle in 2017. Alongside such incredible collaborators, she inspires writers through the process of polishing final details and publishing their dream-come-true.

Skinner Self-Publishing Services
Where every "smidgen" makes a difference.
smidgenpress.com / home

facebook.com / skinnerselfpub
twitter.com / happygrammarian
pinterest.com / skinnerselfpub
instagram.com / skinnerselfpub

Made in the USA
Coppell, TX
21 May 2020

26176270R00125